Springer Theses

Recognizing Outstanding Ph.D. Research

Aims and Scope

The series "Springer Theses" brings together a selection of the very best Ph.D. theses from around the world and across the physical sciences. Nominated and endorsed by two recognized specialists, each published volume has been selected for its scientific excellence and the high impact of its contents for the pertinent field of research. For greater accessibility to non-specialists, the published versions include an extended introduction, as well as a foreword by the student's supervisor explaining the special relevance of the work for the field. As a whole, the series will provide a valuable resource both for newcomers to the research fields described, and for other scientists seeking detailed background information on special questions. Finally, it provides an accredited documentation of the valuable contributions made by today's younger generation of scientists.

Theses are accepted into the series by invited nomination only and must fulfill all of the following criteria

- They must be written in good English.
- The topic should fall within the confines of Chemistry, Physics, Earth Sciences, Engineering and related interdisciplinary fields such as Materials, Nanoscience, Chemical Engineering, Complex Systems and Biophysics.
- The work reported in the thesis must represent a significant scientific advance.
- If the thesis includes previously published material, permission to reproduce this must be gained from the respective copyright holder.
- They must have been examined and passed during the 12 months prior to nomination.
- Each thesis should include a foreword by the supervisor outlining the significance of its content.
- The theses should have a clearly defined structure including an introduction accessible to scientists not expert in that particular field.

More information about this series at http://www.springer.com/series/8790

Nils Braun

Combinatorial Kalman Filter and High Level Trigger Reconstruction for the Belle II Experiment

Doctoral Thesis accepted by
the Karlsruhe Institute of Technology,
Karlsruhe, Germany

Springer

Author
Dr. Nils Braun
ETP
Karlsruhe Institute of Technology
Karlsruhe, Germany

Supervisor
Prof. Florian Bernlochner
ETP
Karlsruhe Institute of Technology
Karlsruhe, Germany

ISSN 2190-5053 ISSN 2190-5061 (electronic)
Springer Theses
ISBN 978-3-030-24996-0 ISBN 978-3-030-24997-7 (eBook)
https://doi.org/10.1007/978-3-030-24997-7

This Springer imprint is published by the registered company Springer Nature Switzerland AG
The registered company address is: Gewerbestrasse 11, 6330 Cham, Switzerland

Supervisor's Foreword

We have entered an era in particle physics in which experimentation is led by computer programs and complex algorithms. Today's experiments have reached a complexity and intricacy such that no other way of analyzing the large streams of data—collected by instruments the size of houses and covering the area of soccer fields in some cases—can be imagined. It is fair to say that neither the discovery of the mechanism of charge-parity violation nor the confirmation of the existence of the Higgs boson, both feats that were rewarded with Nobel prizes, would have been possible without the use of state-of-the-art software and complex algorithms. And it is also reasonable to believe that such tools will play a crucial role in future discoveries. In this thesis, Dr. Nils Braun documents something one could consider the backbone of an experiment: During his Ph.D., he carried out work at Belle II and worked extensively on algorithms that reconstruct trajectories of charged particles traversing the detector. In addition, he made crucial contributions to the so-called online computing, which uses large computing farms to partially reconstruct collision events to decide which of them are of interest to physicists and should be stored to disk. At the time of writing, the Belle II experiment has just started to record electron and positron collisions produced by the SuperKEKB accelerator facility north of Tokyo in Japan. First preliminary physics results are expected this summer, and all of them will make use of the work carried out in this thesis.

Let me describe a typical collision at Belle II: Electrons and positrons are accelerated to nearly the speed of light and brought into colliding paths at the center of the Belle II detector. The detector has the size of a four-story building (with another building of similar size, shielded from the harsh radiation environment such collisions create, to read out the individual sub-detectors). During some collisions, a special type of quantum annihilation occurs, which produces two heavy Beauty quarks. These very unstable elementary particles decay after a very short lifetime of about 1.5 ps into a number of charged and neutral particles and these decay remnants allow us to study and understand the properties of our universe. Do new heavy fundamental particles exist? Is there a connection between the elusive Dark matter and everyday visible matter? In some sense, Belle II is a time machine that allows us to recreate conditions shortly after the Big Bang, when Beauty quarks were abundant and their properties had an impact on the future evolution of the universe. But let us return to Belle II: The decay remnants of both Beauty quarks travel through the detector, and in case of charged particles, three dedicated track reconstruction detectors reconstruct their trajectories. At the very center of the detector and closest to the collision point, a pixelated small sensor with over 7 million individual pixels is located. The volume outside is instrumented by double-sided strip-detectors and a sizable gaseous drift chamber. Both of the outer detectors possess a fast readout, and one crucial aspect of the data-taking scheme at Belle II is that the innermost pixel detector cannot be fully read out for each interesting collision. The information of the two outer detectors thus has to be used to extrapolate to a small region of interest on the pixel detector, which in turn can be read out. If this region of interest is wrong, the recorded information is lost, and with it much of the necessary precision needed to carry out physics measurements at the forefront of our field.

This is where Dr. Nils Braun and his thesis enter the scene: he took up this challenge in his outstanding Ph.D. work. He demonstrated that a combinatorial Kalman filter can be used to carry out such an extrapolation with the necessary precision. His Kalman filter implementation now plays a key role in the sequence of track reconstruction algorithms currently being employed by the Belle II collaboration. His thesis offers a very pedagogical introduction into the topic and offers an excellent starting point for, e.g., track reconstruction algorithms for future experiments. In addition to these crucial contributions, Dr. Braun also played a major part in optimizing the execution time of the online reconstruction code that stores or discards collision events selected by a fast hardware-based trigger. This decision is taken by partially reconstructing the collision event using a large computing farm called the high level trigger. The decision has to be executed within a narrow time window (dictated by the incoming stream of collision data and the limited processing power to select collision events for permanent storage). Dr. Braun's contributions here not only helped to keep the reconstruction time within the allocated limits but also contributed to a new error-proof parallelization implementation, which will stabilize the high level trigger operations in the future. All of this is also outlined in his thesis, providing interesting metrics and supplemental material to optimize such systems.

Thus in summary: This thesis is a must-read, it offers comprehensive and detailed documentation of the Belle II charged particle reconstruction algorithms, and it summarizes crucial parts of the online computing system.

Nils, it was a true privilege to work with you and I wish you the very best for your future! No matter what you will end up doing, you will be outstanding at it.

Sincerely

Karlsruhe, Germany
July 2019

Prof. Florian Bernlochner

Acknowledgements

I always knew very well that there were many people out there I could rely on. I would like to express my sincere thanks to all of them, because without them this work would not have been possible.

My first thanks go to Prof. Dr. Michael Feindt for taking over the main lecture and Prof. Dr. Florian Bernlochner for taking over the co-lecture. Michael has given me a lot of freedom in my choice of topics since I wrote my Master's thesis with him at our institute. I do not want to miss the insights into the industry of data science—also besides science—which I got through him. I would like to thank Florian very much for the always good and personal support. Not only the technical advice but also the help with the organizational tasks helped me a lot. His motivating nature and his many new ideas have brought me and our whole group forward.

Track reconstruction and software development are above all teamwork, and our former Convener for Tracking, Dr. Martin Heck, was not only a good leader of the tracking group but has also brought our local working group in Karlsruhe forward with ever new ideas and improvements. His wealth of knowledge and experience was an enrichment for me. I want to thank him for the many years of collaboration and support. My sincere thanks go to Dr. Thomas Hauth. He was my supervisor for large parts of the thesis and my first help in case of problems. I would like to thank him for his competent help, his support despite the great geographical distance, the new ideas we had in the many discussions, and his very open and friendly nature. I could not have wished for a better supervisor.

Since my time as a master student, I have felt very comfortable in our working group and this is mainly due to my colleagues. I would like to highlight my (former) office colleagues Christian, Moritz, and Markus, with whom I was not only able to discuss common topics but also common problems and had a lot of fun. I would also like to thank everyone who proofread my work. My big thanks also go to the admin team of the institute, who managed to maintain a very smooth operation despite many external problems.

Without the work of several open source developers of countless Python packages, the good questions and answers on many blogs and forums on the Internet (especially Stack Overflow) and the large and helpful community of software developers, this work would not have been possible. My thanks also go to all of them. I would also like to thank the many other developers in the Belle II collaboration, especially Martin Ritter, who answered many of my questions.

Finally, I would also like to thank my family, my friends, and my girlfriend Sophie. They have accompanied me throughout my studies and my doctoral thesis and have helped me with words and deeds.

Contents

Chapter 1
Introduction

Since people have first observed their surroundings, they have been trying to grasp the inner structures that drive the world around them. Mankind has developed complex theories that describe and predict the universe—from the macroscopic theories of galaxy cluster formations to the microscopic theory of particle physics. One of the most successful results of this cross-generational effort is the Standard Model of particle physics, which describes the matter and its interactions through forces discovered so far.

The current theory of the Standard Model is however known to not answer all open questions of particle physics. The measured imbalance between the observed amount of matter and antimatter, the missing explanation for dark matter and dark energy or the incompatible theories of general relativity and particle physics are just some of the pending issues of the model. Large experiments were invented and run to solve those mysteries. Ever-growing international collaborations with experts from multiple fields work on precise detectors, huge accelerator complexes, complex software and difficult data processing to understand the inner secrets of particle physics.

One of the most promising new experiments at the high-precision frontier is the Belle II experiment [1] currently being commissioned at the SuperKEKB accelerator in Tsukuba, Japan. The Belle II detector is a general-purpose high-energy particle physics detector built around the collision point of the e^-e^+ produced by the accelerator. The intermediate $\Upsilon(4S)$ resonance decaying into only two B mesons offers a perfect laboratory for a variety of precision measurements of the Standard Model and searches for new physics, e.g. meson spectroscopy, searches for rare decays, lepton-flavour violations or dark-matter searches. For performing these measurements with a high precision, large data samples are needed. Therefore, the SuperKEKB accelerator was upgraded to be able to deliver an unprecedented, world-record breaking instantaneous luminosity of $\mathcal{L} = 8 \times 10^{35}\,\text{cm}^{-1}\,\text{s}^{-2}$, which is 40 times higher than was achieved at its predecessor, KEKB. Due to the large increase in the collision rate

N. Braun, *Combinatorial Kalman Filter and High Level Trigger Reconstruction for the Belle II Experiment*, Springer Theses,
https://doi.org/10.1007/978-3-030-24997-7_1

and the demanding needs of a high-precision measurement, the detector hardware and the reconstruction software were upgraded or entirely rebuilt.

The increased luminosity is, however, accompanied with difficult challenges to hardware and software. As the large amount of recorded data cannot be stored to disk, a fast and reliable online trigger system is used to suppress physically uninteresting decays. Strong processing time requirements are imposed on the software part of this trigger—the high level trigger (HLT)—as well as constraints on the output rate and the efficiencies for important decay channels. For this thesis, the software used for the HLT decision was developed from scratch. The first studies shown in this work confirm whether it is feasible to run the offline reconstruction also on the online software trigger, which has major benefits compared to using a different, dedicated reconstruction. A replacement for the legacy multiprocessing framework based on the open-source library ØMQ with beneficial new features was implemented in this thesis for the first time.

A basic ingredient for many reconstruction steps is a precise knowledge of the event collision time T_0, which cannot be deduced from the accelerator revolution frequency alone, as the readout windows of the detectors are too large. The estimation of this value needs to be performed as exact and reliable as possible, since many detector reconstructions depend on it. In this work, a combined time extraction method based on the drift time information of the central drift chamber (CDC) was developed and studied.

One of the most important steps in both the online and offline event reconstruction is the search for the charged decay products using the tracking detectors. Tracking is the only reconstruction part able to extract the parameters of all charged final state particles with a high precision. The increased instantaneous luminosity comes with a drastically enlarged beam-induced background, especially in the innermost silicon-based tracking detectors, which imposes major challenges to the tracking. To address these issues, a combinatorial Kalman filter (CKF) was introduced into the Belle II software through the work of this thesis. It enables a wide range of applications during track finding and is used for attachment of hits and tracks in the pixel detector (PXD) and the strip-vertex detector (SVD). The obtained improvements to the vertex resolution are crucial in allowing a larger range of physics channels to be studied and a higher precision in all measurements.

This chapter closes with a more student-friendly introduction into the topics of this thesis, which was only added to this version of the text. The Belle II detector is introduced in Chap. 2 with a focus on the tracking subdetectors and the trigger setup required for this thesis. The foundations behind the development of tracking algorithms and the concept applied at Belle II are described in Chap. 3. In Chap. 4, the detailed processing time measurements for the HLT reconstruction as well as the developed trigger setup are discussed together with the new implementation for the multiprocessing framework. The first-developed precise event time determination required for a successful data acquisition is described in Chap. 5 together with its performance. The combinatorial Kalman filter developed during this work is presented in Chap. 6. Its large beneficial impact on the resolution of the reconstruction and other important figures of merit as well as its performance on the first recorded data is outlined in the chapter. The thesis ends with conclusions on the performed work in Chap. 7.

1.1 General Introduction into Belle II and This Work

In the original form of this thesis, Chaps. 2 and 3 only include the foundations necessary for this work without giving a broader overview on the topic. This section should give readers not especially familiar with the Belle II experiment, B-physics or the topics of this thesis some more general introduction and additional references for further readings.

As most physics experiments, Belle II is built to explore a particular field of physics by measuring quantities which can be predicted by theory. The theories used in high-energy particle physics are summarized in the Standard Model (SM) of particle physics. In this model, all matter is built of point-like elementary particles grouped into six quarks (up u, down d, strange s, charm c, beauty or bottom b, top t) and six leptons (electron e, muon μ and tau τ with accompanying neutrinos ν_e, ν_μ, ν_τ). Those particles interact by exchanging force-carrying particles (photon γ, W- and Z-bosons W, Z and gluon g) and acquire masses via the Higgs-mechanism (represented by the Higgs boson H). Due to the interactions, bound-states like the proton or neutron can form out of the elementary particles and these bound states (mostly also called particles) as well as the elementary particles, can decay and convert into particles of different kind. Good introductions into the topic of the Standard Model are given in [2, 3].

According to the SM, those interactions are not random, but follow certain symmetries. The two symmetries most interesting for the field of B-physics—the main topic studied at Belle II—is the so called SU(2)-symmetry of the W- and Z-bosons and the approximate symmetry of the particle generations, which groups quarks and leptons in 3 families, each one being a heavier version of the generation before (except for neutrinos, which are assumed to be massless in the SM). A lot more information on symmetries and their influence on particle physics can be found in many textbooks, e.g. [4]. The interaction with the W-boson allows the particles to convert from one generation into another, which opens a wide variety of processes to study—the so called field of flavor physics [5, 6]. The b-quark being the heaviest quark which still forms bound states (the top quark decays too fast for a bound state measurable as a resonance to form) has the largest amount of possible decay channels and is therefore a good candidate for studies. The SuperKEKB accelerator together with the Belle II detector are a B-factory, where B-mesons—as the bound state of a b-quark with another quark are called—are generated in large amounts to study their decays.

The basic principle of most high-energy particle physics experiments including Belle II is to collide relatively well-understood particles like electrons or protons with a known energy to produce other particles, which decay according to the rules of the theory. In most cases, this implies several decay steps as the primary high-energetic and most-interesting process happens on time-scales immeasurable for any readout. The final decay products are reconstructed by measuring their properties like energy, momentum and charge using large detector systems built around the collision point. Using the properties of the final decay products, the decay chain is reconstructed back to the initial collision until all (or most) of the processes are understood and categorized (for more information on the methods used for this decay chain reconstruction, see e.g. [7]). The extracted features of the processes can be compared to the theory prediction to find deviations or hints for new physics beyond the Standard Model.

As particle physics is described with the theory of quantum physics [8, 9], only the probability for a certain reaction can be determined, not the precise outcome of every collision. This fact together with the uncertainties involved in every measurement and the rareness of the interesting processes to happen implies the need for a large dataset of B-meson decays under study. Therefore KEKB, which accelerates the electron and positron beam that is used for creating the B-mesons, is upgraded to SuperKEKB to reach a rate of collisions 40 times higher than its predecessor KEKB.

Using this large dataset of B-meson decays as well as other reactions, many physics analyses will be performed. One of the most well-known studies is the analysis of CP-violation in the B-meson system. The mechanism of the generation-change of the quarks and leptons due to W-boson exchange—often described by the so called CKM-matrix—is not symmetric for particles and anti-particles. First hints to this symmetry violation—the P-violation—were discovered in 1956 by Chien-Shiung Wu and were since then studied in great detail [10]. The study of this violation might give indications on physics beyond the Standard Model as well as explain the origin of the observed asymmetry between matter and anti-matter in the universe. Other interesting studies include the search for dark matter candidates, meson-spectroscopy and studies of rare processes [11]. B-mesons are produced at the Belle II experiment via the decays of the $\Upsilon(4S)$ resonance. This resonance is a bound state of two b-quarks originating from a colliding e^+e^- pair. The energy of the collision is tuned in a way to reach a high probability for producing the $\Upsilon(4S)$, as it decays into a pair of B-mesons in more than 95% of the cases. Figure 3.1 in Chap. 3 shows a possible decay chain of the two B-mesons. The shown final-state particles π, ℓ and γ have a life time long enough to be measurable in the Belle II detector. Charged particles are identified in the tracking subsystem and particles with electromagnetic interactions in the electromagnetic calorimeter. Also the p, n and the K count to the group of final-state particles. All other particles or resonances must be identified via their decay products. In a later analysis of the shown $B^+ \to \ell^+\nu\gamma$ decay, a possibly interesting quantity is the decay rate, which depends on one of the CKM-matrix elements (V_{ub} in this case).

With the planned data sample of several billion recorded B-meson decays and orders of magnitude more collisions it is impossible to do any analysis or reconstruction of the data manually. Large efforts are therefore undertaken to write software performing the reconstruction of the data. The tasks of the software include reconstructing the final state particles from the recorded detector measurements, combining them until the primary decay is reached as well as analyzing the extracted features such as event rates or measured particle properties. Typically, this reconstruction starts with the raw digitized detector data directly from the detector and consists of many substeps—first reconstructing particles in each sub-system independently before combining all information to acquire knowledge about all measurable particles produced in a collision. The goal is to know both their kinematic properties such as momentum and energy as well as their particle type and charge. Different sub-detectors are optimized for certain parts of the particle spectrum or to extract certain properties, e.g. the tracking system gives information about the kinematics whereas the particle identification system is used to distinguish between different particle types. This reconstruction does not only happen on the recorded data (so called offline reconstruction), but also needs to be performed directly after each collision to determine if the event is worth storing to disk (the so called trigger stage or online reconstruction). One part of the online trigger system is the software-based high-level trigger—in contrast to the FPGA-based Level 1 trigger. Although the accelerator is tuned to produce $\Upsilon(4S)$ resonances with a large rate, most of the collisions will generate less interesting topologies such as Bhabha decays. For example, as events with B-mesons produce more than four charged final-state particles, discriminating events with only two charged particles can help to reduce the rate of uninteresting Bhabha events. For this however, a reconstruction and interpretation of the event during the data taking is needed. The large increase in the data rates compared to the former Belle experiment imply the need for a faster reconstruction as well as a very robust data transportation and filtering system. The following chapters give an introduction into these specific topics on top of the general references in this section.

References

1. Abe T, Adachi I, Adamczyk K, Ahn S et al (Nov 2010) Belle II technical design report. arXiv:1011.0352
2. Griffiths DJ (2008) Introduction to elementary particles, 2nd rev version. Physics textbook. Wiley, New York. https://cds.cern.ch/record/111880
3. Halzen F, Martin AD (1984) Quarks and leptons: an introductory course in modern particle physics
4. Schwichtenberg J (2018) Physics from symmetry. Undergraduate lecture notes in physics. Springer International Publishing, Cham. http://www.springer.com/de/book/9783319666303, https://doi.org/10.1007/978-3-319-66631-0
5. Grossman Y (2014) Introduction to flavor physics. In: Flavianet school on flavour physics karlsruhe, Germany, Sep 7–18, 2009, pp 111–144. arXiv:1006.3534 [hep-ph], https://doi.org/10.5170/CERN-2010-002.111, https://doi.org/10.5170/CERN-2014-003.73
6. Blanke M (2017) Introduction to flavour physics and CP violation. CERN Yellow Rep School Proc, vol 1705, pp 71–100. arXiv:1704.03753 [hep-ph], https://doi.org/10.23730/CYRSP-2017-005.71
7. Bevan AJ, Golob B, Mannel T, Prell S et al (2014) The physics of the B factories. Eur Phys J C 74(11):3026. arXiv:1406.6311, http://link.springer.com/10.1140/epjc/s10052-014-3026-9, https://doi.org/10.1140/epjc/s10052-014-3026-9

8. Sakurai JJ (1994) Modern quantum mechanics; rev. edn. Addison-Wesley, Reading, MA. https://cds.cern.ch/record/1167961
9. Griffiths DJ (2004) Introduction to quantum mechanics, 2nd edn. Pearson Prentice Hall. http://www.amazon.com/exec/obidos/redirect?tag=citeulike07-20&path=ASIN/0131118927
10. Bigi II, Sanda AI (2009) CP violation, 2000. [Camb Monogr Part Phys Nucl Phys Cosmol 9,1]
11. Belle-II, Altmannshofer W et al (2018) The Belle II physics book. arXiv:1808.10567 [hep-ex]

Chapter 2
Experimental Setup

The task of the Belle II experiment is to search new physics in particle decays at unprecedented high precision. The following chapter summarizes the basic concepts how this task is performed and which techniques are used for this. The information given in this chapter is described in more detail in various other places. A reference with a general overview is included in each section.

As a first step in analyzing particle decays, resonances which decay into those particles have to be produced artificially. This is done with the help of the large accelerator complex SuperKEKB, which is described in Sect. 2.1.1. Afterwards in Sect. 2.1.2, the detector which is used to examine the final state particles of the decays is depicted, with a more detailed description of the tracking subdetectors. Modern particle physics would be impossible without the development of fast and reliable reconstruction software to reconstruct and analyze the recorded data. The online and offline software system is illustrated in Sect. 2.2 and Sect. 2.3 respectively. The chapter is completed with a short overview on the current status of the Belle II experiment up to the point of writing of this thesis and the future time schedule in Sect. 2.4.

2.1 From Accelerated Electrons to Final State Particles

Both Belle II and the SuperKEKB collider are an extended upgrade of their predecessors. The previous Belle experiment located at the KEKB B-factory successfully collected data between 1999 and 2010 which lead to important discoveries in the area of CP-violation, mesons spectroscopy or B-meson decays [1]. More information on the previous Belle experiment can be found in Refs [2, 3]. The following sections will explain the newer updated versions and only compare to the former experiment where necessary. Most information in this sections is taken from the Belle II technical design report [4].

N. Braun, *Combinatorial Kalman Filter and High Level Trigger Reconstruction for the Belle II Experiment*, Springer Theses,
https://doi.org/10.1007/978-3-030-24997-7_2

2.1.1 Accelerator Complex at KEK

The SuperKEKB accelerator located at the High Energy Accelerator Research Organization in Tsukuba, Japan is a so-called *B-factory*. It is an asymmetric electron positron collider operating at a center of mass energy of $\sqrt{s} = 10.58\,\text{GeV}$, which is slightly above the $\Upsilon(4S)$ resonance. The $\Upsilon(4S)$ resonance is an excited bottomonium state with an energy above the mass of two B mesons. The produced resonance decays in more than 96% of the cases exactly in two B mesons which are produced nearly at rest in the center of mass system (CMS). To allow a measurement of the drift distance of the short-living B mesons, which is a necessary ingredient for CP violation measurements, the center of mass system is boosted with respect to the laboratory system by using asymmetric energies for the electrons (HER, high energy ring of 7.0 GeV) and the positrons (LER, low energy ring of 4.0 GeV).

An overview of the whole accelerator complex is shown in Fig. 2.1. The electron beam is produced in bunches with a photocathode radio frequency gun in the pre-injector at the beginning of the linear accelerator (LINAC) and accelerated up to 4 GeV. Every second bunch of produced electrons hits a tungsten target located in the LINAC to generate positrons which are extracted in a capture section afterwards. To shrink the high emittance and the energy spread of the positrons after their generation, a damping ring operating at 1 GeV is used. Electrons and positrons are accelerated to their final energies of 7 and 4 GeV in the second half of the LINAC and are then injected into the main storage ring.

The 3 km long storage ring can hold approximately 2500 bunches of particles which collide at the interaction point (IP) inside the Belle II detector every 4 ns. Using the nano-beam scheme proposed for the Super B factory in Italy [6], a record-breaking luminosity of $\mathcal{L} = 8 \times 10^{35}\,\text{cm}^{-1}\,\text{s}^{-2}$ can potentially be reached. To reach this luminosity, the vertical beta function β_y^* must be squeezed to values as low as

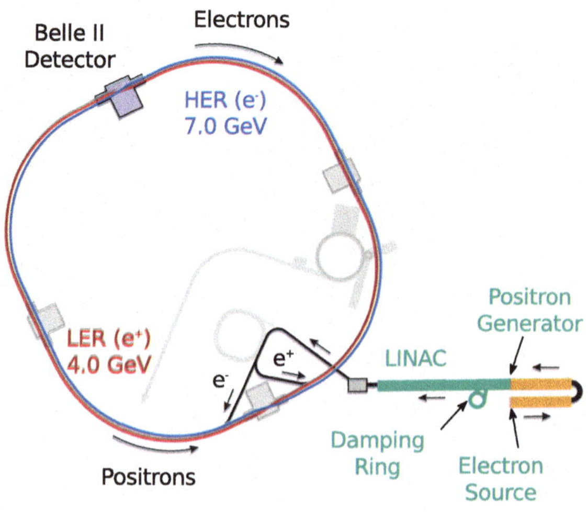

Fig. 2.1 Schematic drawing of the SuperKEKB accelerator complex which generates the electron and positron beam colliding at the interaction point inside the Belle II detector. Adapted from [5]

270 μm. This is only possible by minimizing the effective longitudinal beam size, because larger longitudinal sizes would limit the effective minimum value of β_y^* due to the hourglass effect. A small effective longitudinal size is achieved by a large horizontal crossing angle ($\phi = 41.5$ mrad) and a small horizontal beam size ($\sigma_x^* = 7.75$ μm), which leads to the very small vertical beam size of $\sigma_y^* = 59$ nm. This vertical beam size in the nanometer range gives the scheme its name.

2.1.2 Detector Design

The Belle II detector located around the IP of the SuperKEKB collider is supposed to measure the properties of all final state particles produced in the e^+e^- collisions, except for the non-interacting neutrinos. This includes charged as well as neutral particles, so different strategies are applied. Like in a typical high-energy particle physics detector, different layers of subdetectors each with a different tasks, are arranged around the interaction region. A schematic drawing of the detector in its final form is shown in Fig. 2.2. The different extension levels are discussed in Sect. 2.4.

The detector consists of the following subparts, beginning with the innermost subdetector.

- The properties of charged particles are measured in a tracking system consisting of the pixel detector (PXD), the strip vertex detector (SVD) and the central drift chamber (CDC). PXD and SVD are grouped together as the vertex detectors (VXD). A superconducting solenoid produces a nearly homogeneous magnetic field along the beam direction of 1.5 T which forces the particles on a bent trajectory.
- For distinguishing between π and K particles, a time-of-propagation counter (TOP) in the barrel and an proximity-focusing aerogel ring imaging Čerenkov detector (ARICH) in the forward endcap region are installed.
- Since the tracking systems can only measure charged particles, an electromagnetic calorimeter (ECL) is needed to measure the energy deposition of the neutral pho-

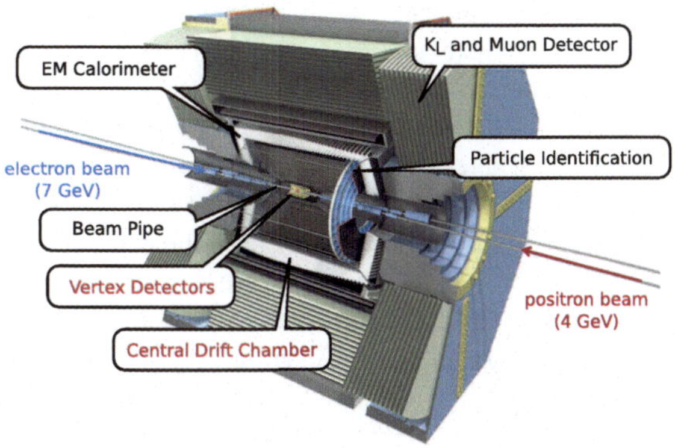

Fig. 2.2 Schematic view of the Belle II detector in its planned final status. The different subdetectors are labeled. The tracking system described in more detail is highlighted. Adapted from [4]

tons using scintillating CsI(Tl) crystals. Neutral π particles are measured via their decay into photons. By measuring the energy and the radiative energy deposition caused by bremsstrahlung for electrons and positrons, additional information is gained for particles which were already reconstructed by the tracking system.

- To identify K_L and μ particles which only interacted weakly with the detector material, the K_L and muon detector (KLM) is used. It consists of glass-electrode resistive plate chambers (barrel) and scintillator strips (end-caps) which are installed in the return yoke of the solenoid.

The CDC, the ECL as well as the TOP all measure the time of the energy depositions. Together, they can be used for determining the global interaction time of the recorded event, which is discussed in Chap. 5 in more detail.

Most final state particles produced in the decays under study—namely $e^{\pm}$, $\mu^{\pm}$, γ, $\pi^{\pm}$, π^{0}, p, n, $K^0_{L/S}$ and $K^{\pm}$—can be observed in the detector and their properties can be measured. Neutrinos are the only particles which do not leave a measurable trace. Information on neutrinos has to be derived by more elaborate reconstruction methods, e.g. a full event reconstruction [7].

The tracking detectors described in the following measure the position and the momentum of charged particles by reconstructing their path through the applied magnetic field. The magnetic field points into the z direction (which is defined to be along the beam axis in the direction of the electron beam) so the measured curvature in the x-y plane (x points horizontally outwards of the accelerator tunnel) gives information on the transverse momentum. Together with the measured angle θ between the track and the z axis, the full momentum can be calculated. Since the position of the particles is measured also, the trajectories can be extrapolated to the IP. There, multiple reconstructed trajectories can be simultaneously fitted to the same vertex. The resulting vertex gives information on which particles stem from the same decay and what properties the intermediate particles have, which is a valuable input for later analyses.

Three special detectors with different tasks and properties are needed to reconstruct the charged particles in a large phase space region. The inner PXD measures the track position with a very good spatial resolution but also produces a lot of data, which needs to be processed fast due to the high collision rate. The outer CDC with its large lever arm can measure the momentum of the particles to the sub-percentage level—although its single spatial measurements have a worse position resolution compared to the vertex detectors. Also, the radiation length in the gaseous detector is very large making it optimal for the momentum range of the particles produced at Belle II, where multiple scattering and material effects play a crucial role. The connection between the outer drift chamber and the inner pixel detectors is given by the SVD. Although not the same as in the pixel detector, the spatial resolution in the SVD is still good enough to perform standalone track finding which increases the finding efficiency especially in the low momentum region, where tracks do not necessarily reach the CDC. However the best resolution for the properties of the particles is only reached, if the information of all detectors is combined. The reconstruction algorithms are described in more detail in Chap. 3.

2.1.2.1 Vertex Detector (VXD)

The measuring principle of the two vertex detectors PXD and SVD is based on using silicon, which is a semiconducting material. As the energy gap between the conduction and the valence band is in the order of a few eV in these materials, charged particles can easily generate electron-hole pairs when flying through the sensor region. These generated free charge carriers can then be measured using external electronics. The number of produced electrons is proportional to the deposited energy. After a conversion to binary format, this measured signal is sent to the detector readout and the trigger.

The PXD is separated into two layers of approximately 8 million pixel sensors and the SVD in four layers of 224.000 silicon strips. The complicated detector design shown in Fig. 2.3 was chosen. It allows to have a constant coverage also in the forward region without any large gaps in the detector acceptance. Still, the material budget is kept as small as possible. Because of the increasing number of sensor strips in the SVD per layer, there is basically no θ angle in the acceptance between 17° and 150° without coverage, as the regions between the sensors mostly do not align in z direction. For the same reason, there is a slight overlap of the sensor ladders (all sensors with the same angle ϕ in the x-y plane) leading to the shown 'windmill' design. This overlap can also be used during the alignment procedure of the detector and gives useful information for the software track reconstruction.

The sensors in the slanted forward part are shaped trapezoidal, the remaining sensors are rectangular shaped. Important design values of the detector are summarized in Table 2.1.

In high energy physics experiments it is common to use very precise pixel sensors at the inner layers for reducing the occupancy, which would occur in strip detectors because of ghost hits. However, the low energies of the SuperKEKB accelerator imposes a severe complication to the PXD design: thick sensors producing large signal yields with direct on-board readout electronics requiring active cooling would imply a large material budget, which would increase multiple scattering. The solution

Table 2.1 Important design values of the two vertex subdetectors as described in [4]

	PXD	SVD
Layers	14,22 mm	38, 80, 104, 135 mm
Number of channels	7680000	224000
Number of sensor planes	40	172
Pitch	50 μm	160–240 μm/ 50–75 μm
Radiation length (approx)	0.19%	0.57%
Thickness	75 μm	300 μm
Integration time	20 μs	300 ns
Expected occupancy	max. 3%	max 10%
Acceptance range θ	17–150°	17–150°

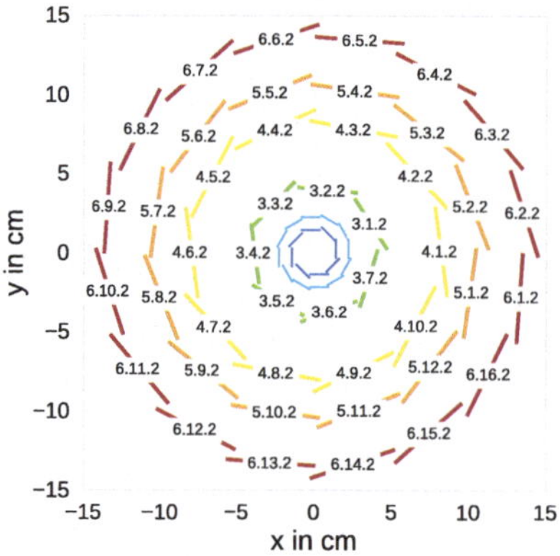

(a) x-y view of the VXD.

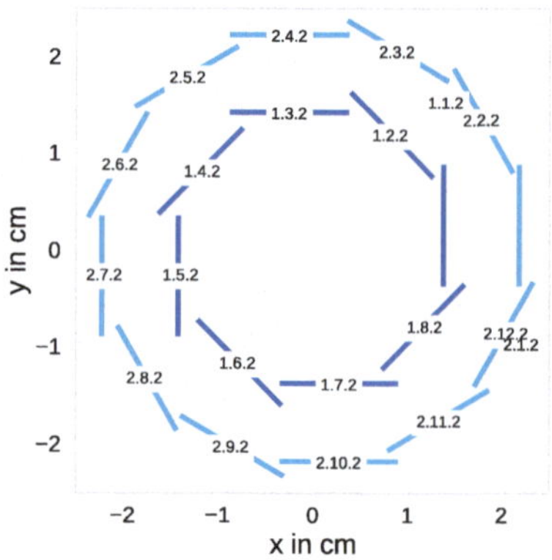

(b) Zoom onto the PXD layers.

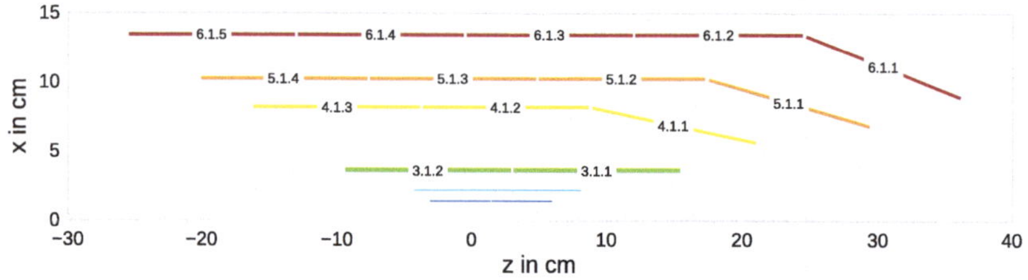

(c) z-x view of the VXD only showing one sensor ladder.

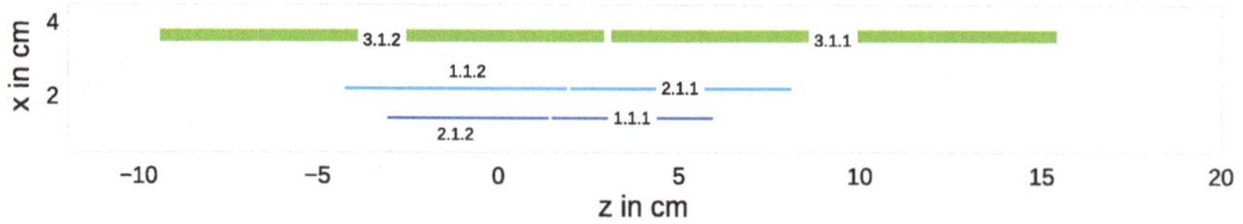

(d) Zoom onto the PXD layers.

(e) Three dimensional view along the z axis.

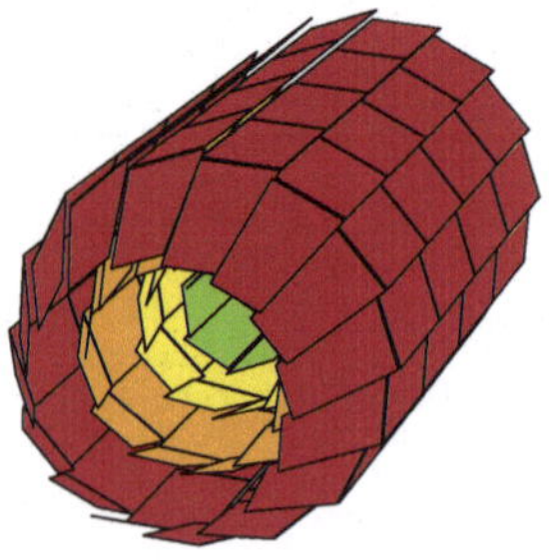

(f) Three dimensional side view.

Fig. 2.3 Geometry of the vertex sensors from different angles. Each line/plane represents one sensor (thickness of the planes exaggerated for visibility). The shown geometry is without misalignment. Not shown are data readout or mechanical structures as well as cooling infrastructure was to choose sensors with the depleted p-channel field-effect transistor (DEPFET) technology, which can be produced very thin due to their internal signal amplification. The DEPFET concept was invented in 1987 by Kemmer and Lutz [8] and is described in detail in [4]. The sensors themselves do not require active cooling because of their low power consumption. The readout electronics is located outside of the acceptance region decreasing the material budget of the PXD sensor layers. This however brings a severe downside: a sensor readout for all channels at the same time is not possible anymore. The pixels are read out in chunks using a rolling shutter scheme, which leads to a total integration time of 20 μs.

This does not only make the PXD unusable for fast trigger decisions, but also increases the occupancy and the data size of a single read out[1]. To cope with the larger data rate, a specialized readout logic for the pixel detector had to be invented, which is described in Sect. 2.2.

Using measurements in the SVD, the very precise position information from the PXD can be combined with the momentum measurement extracted from the CDC. The combined information of multiple tracks can then be used to extract the vertex information.

The SVD uses a similar measurement technology as the PXD which also relies on semi-conducting devices. Using as many channels as in the inner pixel detector would not only be very expensive, but also very hard to read out. Therefore silicon strips with double-sided readout instead of pixels are used, which eliminates these problems. Two measurements on a v–(parallel to the beam) and on a u–(perpendicular to this) strip are combined to produce a space point with three dimensional position information. As a downside, a wrong combination of the strip measurements on the two sides can produce additional so-called *ghost hits* increasing the occupancy of the detector and worsening the signal to background ratio.

The produced electrical signals of the incoming charged particles are sampled with an adjustable frequency. Because of the large number of strips, only a maximal number of six samples of the signal shape can be recorded and transmitted in the readout window of 300 ns. Together with the limited amount of storage of the event data, this poses severe requirements to the trigger jitter and the trigger itself, which are discussed in [9]. Additional to the binary information that a charged particle passed the sensor strip together with the amount of deposited energy, also a time information can be extracted from these samples of the signal shape—although there is a large uncertainty on this value.

Charged particles traversing one of the VXD sensors often deposit energy in more than one pixel or strip in a sensor. The first step after deserializing the raw recorded data from the detector (called digits) is therefore a clusterization, which groups those depositions according to their geometrical proximity and other similarities. Those clusters are the base input for the track reconstruction algorithms. In the following, the terms hit and cluster will be used interchangeably for the VXD.

[1]The probability to have multiple signal events in this time window is small. However, background processes as described below are distributed equally in time decreasing the signal fraction in one PXD readout.

2.1.2.2 Central Drift Chamber (CDC)

The CDC of Belle II is a large multi-wire proportional chamber with a radius of more than 1 m filled with a helium-ethane gas mixture in equal parts. The design mostly follows the global structure of the CDC used for Belle. In total 14336 wires are arranged in 56 layers which are grouped in alternating stereo and axial superlayers. The CDC covers the full acceptance region of 17–150° in θ and 2π in ϕ.

The axial sensor wires together with additional field-shaping wires are tensioned parallel to the beam axis in z direction. For measuring the z component of the momentum without relying on e.g. the arrival time, approximately half of the wires are stereo wires. These wires have mountings that are slightly twisted at both end-caps producing a certain tilt with respect to the z axis. The cells in the innermost superlayer are smaller compared to the rest, as they need to cope with a larger occupancy due to higher beam-induced background (see next section) for smaller radii. Figure 2.5 shows a schematic drawing of the CDC.

Charged particles ionize the helium atoms producing free ions and electrons. Because of their large mass, the ions recombine before they can be measured. The electrons however get accelerated by the applied electric field between the sense and field wires and cascade until a measurable signal is read out at the sense wires. The additional ethane in the gas mixture ensures that the cascading ionization stays localized around the sense wire.

The readout electronic starts a timer on the rising slope of the current at the wire. On a periodical external trigger signal (approximately every 30 ns), this timer is read out. This leads to a measurement of the time between the last external trigger signal and the arrival of the drifting electrons at the wire. After subtraction from the known clock time, the full measured time is given by

$$T_{\text{meas}} = T_0 + T_{\text{flight}} + T_{\text{drift}} + T_{\text{prop}} + T_{\text{walk}}$$

where T_0 is the time of the collision producing the measured track (with respect to the external trigger signal) and T_{flight} denotes the flight time of the particle until it reaches the drift cell. T_{drift} indicates the time the electrons need for drifting, which can be as high as 500 ns and is dependent on the drift length d, T_{prop} is the time the signal needs to propagate on the wire to the readout electronics, which is dependent on the z position along the wire and T_{walk} is the finite rise time of the analog pulse and can normally be neglected. The different time components are shown in Fig. 2.4. How these measurement can be used for the determination of T_0 is discussed in Chap. 5.

The CDC signals are used for the trigger decision as well as for the T_0 determination on the trigger system. During the time the trigger decision is calculated, additional data that may arrive later (because of larger drift times) is recorded and stored in a ring buffer, which is passed along in case the trigger decision is positive.

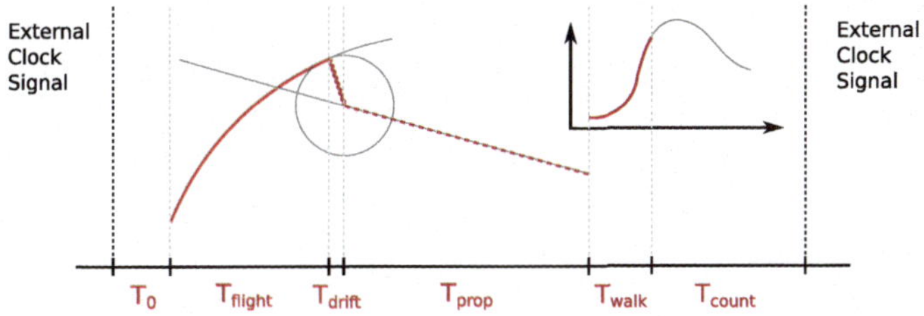

Fig. 2.4 Different components of the measured time for a fired CDC wire. After the particle produced at T_0 traveled to the cell (T_{flight}), the electrons from the ionization drift to the sense wire (T_{drift}). The electrical signal propagates to the readout electronics (T_{prop}) and starts a timer after a rising time (T_{walk}). This timer is read out at the next external clock signal (T_{count}) and can be used to determine $T_{\text{meas}} = T_{\text{clock}} - T_{\text{count}}$ with the known external clock time T_{clock}. The times shown in this figure are not to scale

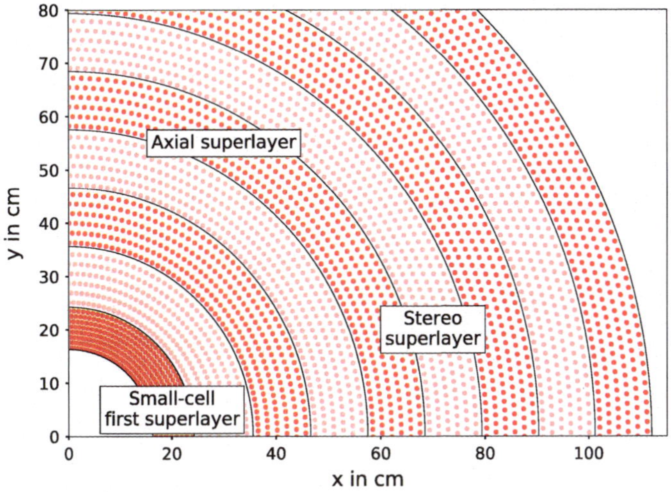

Fig. 2.5 Geometry of the central drift chamber in the x-y plane. As the stereo wires have no defined x-y coordinates, the closest approach to the interaction point for each wire is taken. The geometry is shown without misalignment or any mechanical structures

2.1.3 Beam-Induced Background

The presented devices, especially the tracking detectors, rely on the measurement of the deposited energy of the produced final state particles. Apart from these signal decays under study and the unavoidable but mostly small electronic noise, there are other background processes leading to energy depositions in the detector. In most detectors, the number of background-induced energy depositions is equal or even higher than the number of the ones coming from signal. As the background hits mimic typical signal hits, they pose a severe problem for the event reconstruction.

Most of these background sources are unique to e^+e^- colliders and cannot be seen at proton colliders like the LHC experiments. These processes are amplified with the high beam current and the small beam size at the interaction point at Belle II, so the anticipated record breaking luminosity will worsen the problem compared to the predecessor Belle. During the development of the detector hardware, this was already taken into account at various steps, e.g. with the small-cell drift cells in the inner region of the CDC, the decrease in the boost between the HEP and the LEP beam and the interaction region design [4].

Synchrotron Radiation (SR) Due to the transversal acceleration of the electrons and positrons in the storage ring, the leptons emit synchrotron radiation with high intensity. This is especially the case in the interaction region with its strong focusing magnets. The produced photons have an energy high enough to cause ionization in the detector material of the inner detector layers.

Beam-Gas Scattering The interactions of the accelerated bunches with residual gas in the beam-pipe causes bremsstrahlung and Coulomb scattering. This changes the kinematic properties of the beam particles making them deviate from their nominal values. These disturbed particles hit the walls of the beam pipe or the surrounding magnets producing secondary showers of particles, which can leave measurable traces in the detector.

Touchek Scattering The same effect can be caused by intra-bunch scattering of the beam particles. Since this is proportional to the beam size, this background will play a major role for Belle II. Fortunately, the rate of Touschek scattering also scales with the beam energy E as E^{-3}, so increasing the energy of the LER compared to Belle is a first countermeasure.

Radiative Bhabha Scattering The cross section for Bhabha interactions of the accelerated electrons and positrons is much enhanced compared to the one of signal processes. These radiative processes do not only produce collision events, which must be filtered out by the trigger system, but also produce photons and beam particles with non-nominal kinematic properties. This leads to the same anticipated consequences as described above.

Electron-Positron Pair Production The most severe background for the innermost detector layer are low-momentum particles created in the two-photon process $e^-e^+ \rightarrow e^-e^+e^-e^+$. The number of background energy depositions on just the first PXD layer in one readout can go up to 14,000 hits with the planned luminosity compared to around ten anticipated signal hits.

If not otherwise mentioned, simulations of the beam-induced background will be used through this thesis. Although the simulations are carefully validated with the first measured data, the final background level is still not known at this point and the simulations are only an estimate.

2.2 From Digital Detector Data via the Trigger to the Storage Drives

Although the number of particles in a single bunch is high, not every collision produces an interesting signature—especially not a B meson decay. The collision rate is dominated by elastic Bhabha scattering processes, which are not of interest for the physics studies performed on the Belle II data (see also Chap. 4). As the high number of those physically uninteresting processes cannot be stored or transported and they are also not needed for later analyses, early filter stages on hardware and software are necessary. Additionally, the detector readout of the subdetectors needs to be synchronized to make sure that each device records the data of the same collision event.

The full readout chain is shown in Fig. 2.6. In the following, the parts interesting for this thesis are described in more detail.

2.2.1 Detector Readout and Hardware Trigger

Belle II's online trigger (also called Level 1 or L1 trigger) is mainly built upon two independent trigger lanes, the CDC and the ECL trigger. The different subdetectors produce near real-time trigger signals in the order of 16 ns intervals. All subdetector triggers are combined to a global L1 trigger decision in the global decision logic (GDL) after a fixed latency of 5 µs. If the decision is positive, this information is transferred via the Belle2link connection [11] to the front-end readout boards of all subdetectors (except PXD). Those boards were storing the full event data in a ring buffer during the 5 µs. The hardware triggers and the front-end electronics are implemented using field programmable gate arrays (FPGA). The detector data is then delivered to the common pipelined platform for electronics readout (COPPER). This platform transfers the data eventually to the event builder computers. The information is collected and sent via an ethernet connection to the high level trigger (HLT) running a software reconstruction.

There are different trigger channels planned for the L1 trigger and the final decision on the output rate for each physics process is still under investigation. However, it is already planned to have a total output rate of 20 kHz at the anticipated luminosity of $\mathcal{L} = 8 \times 10^{35}\,\text{cm}^{-1}\,\text{s}^{-2}$. Including a safety margin the rate may be as high as 30 kHz. An overview on the current status of the hardware trigger menu is given in [12].

2.2.2 Online Software Trigger

The decision whether to finally keep the event after the software trigger is based on variables calculated on the event topology and the physical properties of the

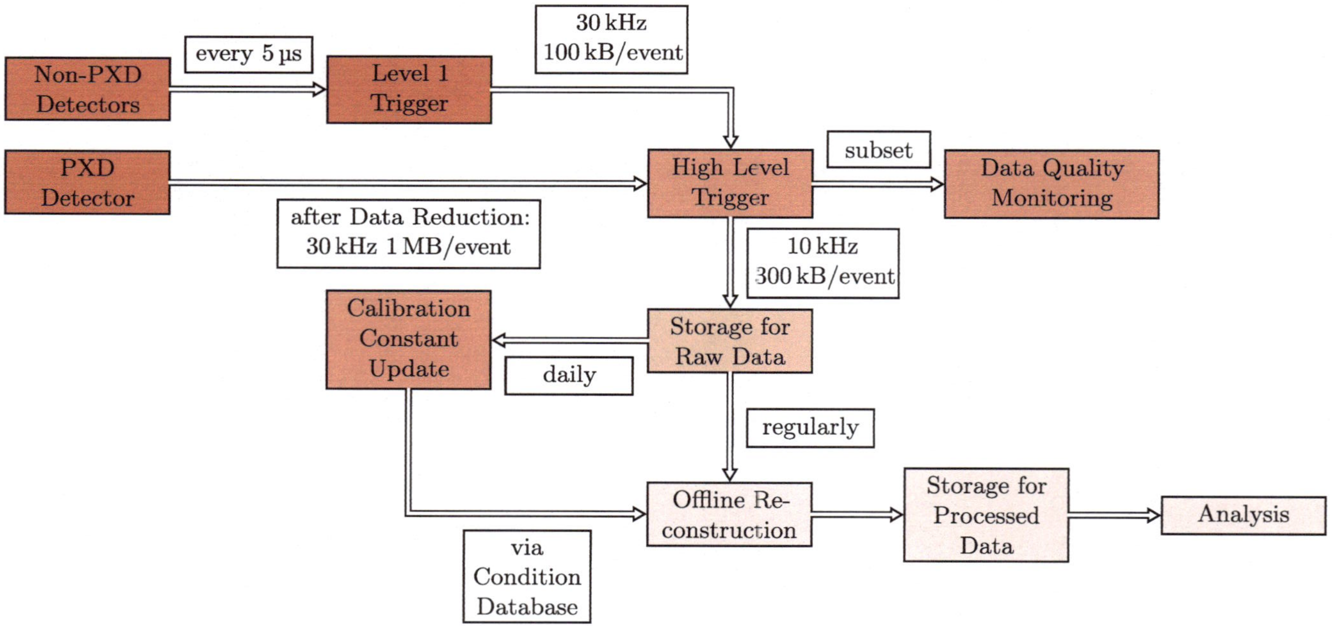

Fig. 2.6 Schematic overview of the data flow from the detector readout followed by the different trigger stages up to the storage devices. The stored data is then reprocessed asynchronously and used for physics analyses. The shown data rates are preliminary and are taken from the current software simulation using the physics rates anticipated in [4, 9, 10]. Specific parts are shown in more detail in Fig. 2.7

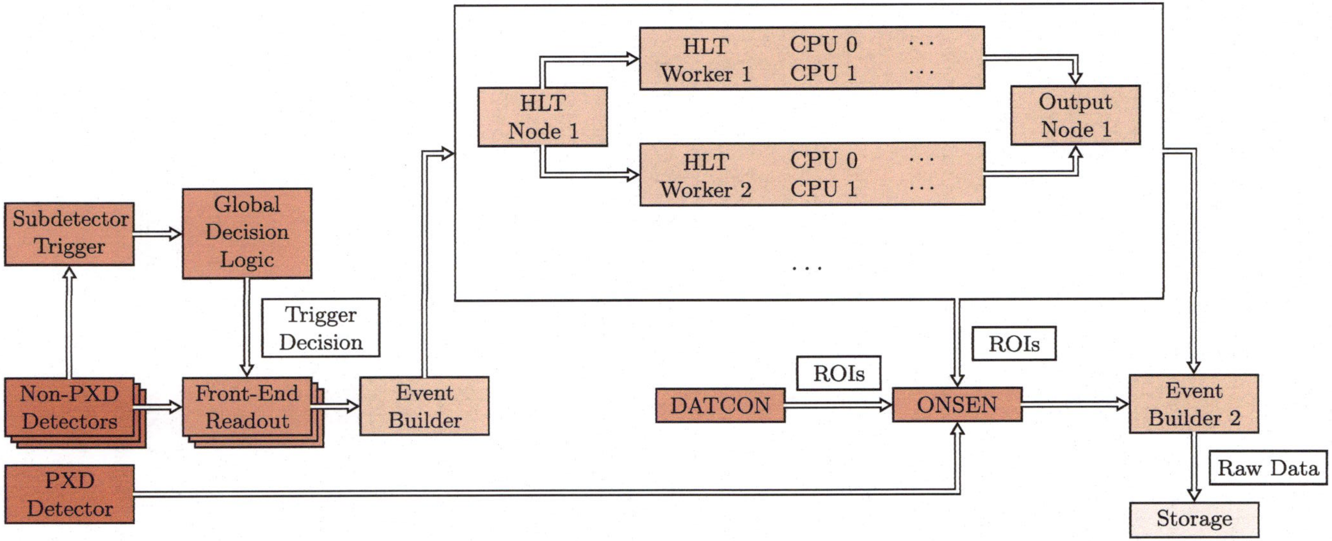

Fig. 2.7 Detailed description of the trigger parts of Fig. 2.6. The non-PXD data is fed through the Level 1 trigger and the HLT to the Event Builder 2. The PXD data is only read out in specific regions of interest (ROIs) given by the HLT and the DATCON. Adapted from [4, 9, 10]

particles. This implies, that the particles are fully reconstructed using the detector data, which is a highly demanding task. This task cannot only be implemented on low-level hardware but needs software running on a server farm very similar to the offline reconstruction described in the next subsection and in Sect. 2.3. The online reconstruction as a faster subset of the full reconstruction is described in more detail in Chap. 4.

The full detector information without the PXD is distributed from the event builder computers over ethernet to the HLT nodes and then eventually to the HLT workers. For the full luminosity run, there are approximately 20 HLT units planned with each of them hosting 16 HLT workers with 20 CPUs. As the online trigger software runs on each of the CPUs of the workers, there are around 6400 processes in parallel calculating the trigger decision. With the expected L1 output rate of 20 kHz (maximum 30 kHz), each process has on average 320 ms (213 ms) for calculating the trigger decision. The HLT should reduce the event rate to approximately 10 kHz. More information on the HLT processing time and rates can be found in Chap. 4.

Apart from the mere binary decision if an event should be stored or is not of interest, the HLT outputs a classification information (so called tag) which can be used to separate the recorded events into different sets according to the underlaying physical process. This tag information is later used to faster skim through the recorded data when looking for a specific process to analyze.

As described above, the integration time of the PXD is very long leading to a large latency and a large amount of hit pixels in a single readout. It is not possible to transfer or store the full data, so a way to reduce the number of hits in a single readout is needed. A very important task of the online reconstruction is therefore to deliver information on where to expect those PXD measurements which are necessary for increasing the resolution of already found tracks. This is done by extrapolating the reconstructed particles into the volume of the PXD. A rectangular-shaped region around the extrapolated positions on each sensor plane, where a particle is expected to travel through the detector (also called region of interest or ROI) is then sent to the readout board of the PXD. The online selection nodes (ONSEN) combine these ROIs with a similar information extracted by a faster but more course FPGA-based ROI extraction (the data concentrator—DATCON) and transfers only the PXD hits in the given regions to the second event builder. The information from the pixel detector together with the remaining subdetectors is then combined and sent to the storage system.

During the recording of the collisions, the data needs to be monitored to assure a constant data quality without degradation in neither the detector systems, the triggers nor the online reconstruction. A defined fraction of triggered events together with the PXD information is not only sent to the storage system, but also to the Express-Reco farm, which executes the default offline reconstruction and extracts a set of monitoring plots, which are shown integrated over multiple events to the control room shifters. Software for automatic deviation detection helps to identify possible failures of hardware and software [13].

2.2.3 *Storage and Offline Reconstruction*

The combined subdetector data of the events, that survive the different trigger stages is stored to disk afterwards. The estimated storage capacity until 2021, where an integrated luminosity of around 20 ab^{-1} will be reached, is around (35.9 ± 9.1) PB [14]. Asynchronously to the data taking, the recorded data is reprocessed using a slower but more precise offline reconstruction running on a large computing cluster. During this reconstruction, information on the detector conditions (alignment and calibration constants) during the data acquisition is taken into account. Those constants are calculated on the data itself by using well-known physics processes with a distinct topology that allows to measure e.g. the displacement of the different detector layers among each other. The process of calibration and alignment is repeated regularly on the recorded data and the result is fed into a database, which is both used for the reprocessing of the data as well as for future online software runs.

The output of the offline reconstruction is a summary of the reconstructed particles suitable for physics analyses known as mDST. Skimming campaigns sort the data according to their HLT tags and other variables calculated on the reconstructed particles to make physics analyses faster.

The reconstructed as well as the raw data is distributed to different computing centers around the world for better accessibility and data backup.

2.3 Belle Analysis Software Framework 2

Software plays a crucial role for the Belle II experiment. It is not only used for taking data and triggering and later for the offline reprocessing and reconstruction, but also during analysis of the data for studies or calculations of branching fractions, discovery of new particles or for outreach and tutorials.

The Belle analysis software framework 2 (basf2) [15, 16] described in this section is used to fulfill all of those requirements. Its basic building blocks are *modules* mostly written in C++ [17] or Python [18], which accomplish a certain task at hand. Those modules are assembled and controlled using *steering* scripts written in Python, which gives the possibility to built complex applications out of the single blocks. The modules are able to use external libraries like ROOT [19], Eigen [20] or EvtGen [21], which are commonly used and well tested in the high energy physics community.

In these steering files a so-called *path* of modules is built, which may for example consist of data input from disk, a reconstruction of the event (which consists of more than one hundred modules) and an output module to store the reconstructed data to disk. The path is then processed by the framework, which handles the loading of the libraries and the consecutive looping over all events from the input. An interface between Python and C++ is implemented using the boost library [22] for the core functionality and PyROOT for user code. The data streaming needed to transfer and store intermediate reconstruction results is done with ROOT's own automated serialization framework based on data member dictionaries created by cling [23]. An integration with jupyter [24] simplifies the creation of complex analyses [25].

2.3.1 Inter-Module, Inter-Process and Inter-Node Communication

As each of the different basf2 modules accomplishes only a certain task, e.g. filter out noise measurements of a certain detector, there is the need for communication between modules. This communication is handled via a common data store, where each module is allowed to read from or write to. Each entry is identified with its name and the type of the stored objects. Using the serialization techniques mentioned above, the data store can be turned into an unstructured byte stream or into structured dictionaries. The data may be written to disk or transferred to a different location via inter-process or inter-node communication.

For online reconstruction on the HLT worker nodes as well as for local development and processing of data, the possibility to run basf2 processes in parallel is needed. Accepting a larger memory consumption, basf2 does only include a multi-process and no multi-thread processing, because this would complicate the development of the software significantly. The multi-core event processor implemented in the framework [16, 26] exploits two ring buffers for communication between an input, multiple worker and an output process. All processes run a path of basf2 modules similar to the single-core mode. The different processes communicate by sending the full data store content as a bytestream via the ring buffers. The process is fully transparent to the user as the modules in the path are automatically split up into input, worker and output parts. The full schema is shown in Fig. 2.8. Chapter 4 describes a different solution developed during this thesis.

The communication between nodes, e.g. between the event builders and the HLT nodes or the HLT nodes and the associated workers is also done by streaming the data store—this time via the TCP protocol. More information on this topic can be found in [26].

2.3.2 Monte Carlo Simulation

The description of the software up to now was about the online and offline reconstruction of the data taken by the experiment. In contrast, most of the results of this thesis are based on studies using simulated events—so called Monte Carlo (MC). Simulated events resembling the later expected recorded data has many advantages. Basically, it is available infinitely and the physics process leading to the data can be chosen at will. Additionally, the truth information is always present making comparisons with the reconstructed information possible.

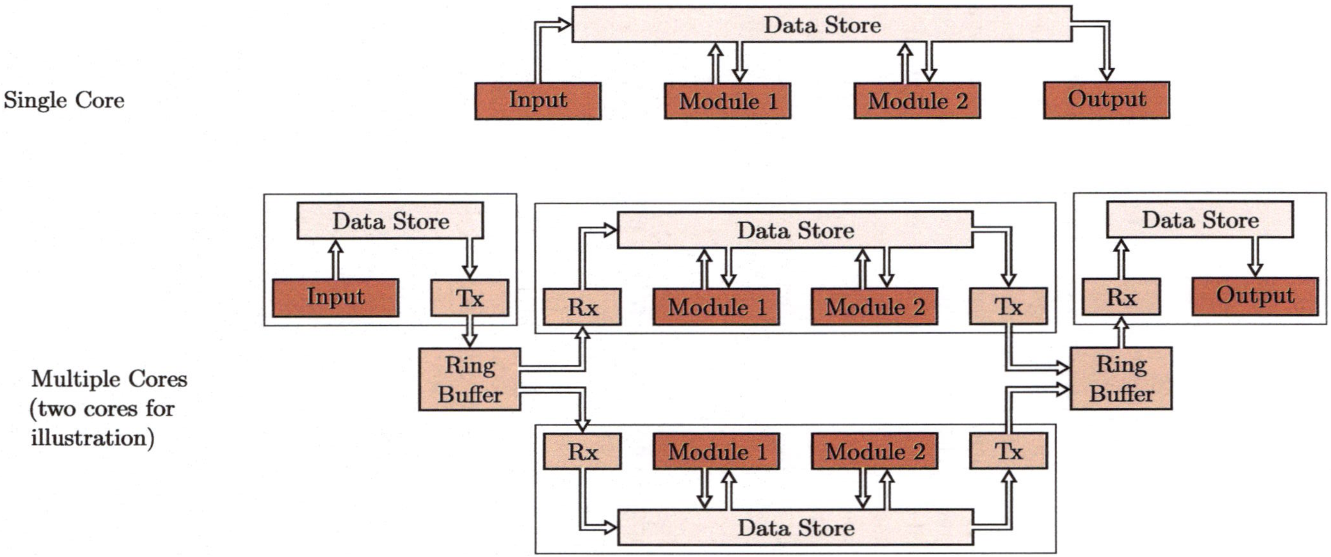

Fig. 2.8 Comparison of the single- and multi-core event processing handled by the framework. In both cases, each process (indicated by the black rectangle in the multi-core case) handles a part of the basf2 module path. The communication is done by streaming the full data store into a ring buffer shared between the processes. The streaming is performed by the transmission (Tx) and receive (Rx) modules

The process of generating MC events which resemble the later detector recording is quite complex and only described briefly here. More information can be found e.g. in [27].

As a first step, a hard scattering process between an electron and a positron is simulated with e.g. the EvtGen [21] package. The produced particles are further hadronized into mesons and baryons using Pythia [28]. Geant4 [29] is used to transverse these mostly long-living final state particles through the simulated detector material. At this step, all known interactions between the particles and the detector material are taken into account. For this, a precise knowledge of the physical processes as well as the detector geometry is needed, which is validated using recorded data. This step produces the energy depositions in the sensor planes, which are then artificially digitized into digital detector signals taking into account the readout properties of the subdetectors. The event is overlayed with simulated—or at a later stage also recorded—background hits. To resemble the real detector data even more, the simulated digital detector hits can be transformed into the low-level raw format, which has the same format as the bytestream of data coming directly from the front-end electronics (a process known as packing). Following this stage, the normal reconstruction of data can be executed on the MC simulation without any changes. The truth information is kept separately, but can still be accessed for analysis of the reconstruction results, e.g. by the track matching described in Sect. 3.2.3.

2.4 Time Schedule and Performed Tests

Before the physics data acquisition of Belle II can start, excessive testing of the detectors and the software has to be performed to achieve the aspired precision and performance. Parts of the subdetector systems were already tested during their assembly, but only a full detector test with the complete readout chain with realistic conditions will show where possible problems need to be solved. These tests started with Phase 2 of the Belle II experiment in early 2018, when most of the detector systems were integrated and tested together. The first tests on cosmic ray data and low-luminosity collisions showed that the data acquisition, the readout and the trigger system with the online reconstruction on the high level trigger worked mostly as anticipated (see e.g. [30]). As the background conditions were only simulated up to this point, dedicated measuring devices were included near the interaction region to compare MC simulations with the recorded data. Additionally, the full vertex detector was replaced by just a single slice of low-quality sensors in case of unexpected high background radiation which may destroy the detector. The data taken in this period is mainly for commissioning the different detector systems and the data processing including the calibration and alignment, the data storage and offline data reprocessing. A description of the Phase 2 setup will be given in Chap. 6.

Phase 3 will start the full physics data taking period with the planned instantaneous luminosity of $\mathcal{L} = 8 \times 10^{35}\,\text{cm}^{-1}\,\text{s}^{-2}$ after a ramp up employing the full detector including the VXD. It will start in early 2019. Most studies performed during the work of this thesis are done for the detector and background conditions of Phase 3 and the following data taking periods. When possible, the algorithms were also tested with the data recorded in Phase 2. During the first years of data acquisition, it is planned to reach the integrated luminosity of the predecessor Belle. After multiple years of data taking, a dataset 40 times larger than the present one will be cumulated decreasing the statistic and, because of the better detector, also the systematic uncertainty on many interesting physics channels.

References

1. Brodzicka J, Browder T, Chang P, Eidelman S et al (2012) Physics Achievements from the Belle Experiment," *Prog Theor Exp Phys* 2012(1), 4D001–0. arXiv:1212.5342, https://doi.org/10.1093/ptep/pts072
2. Abashian A, Gotow K, Morgan N, Piilonen L et al (2002) The Belle detector. Nucl Instrum Methods Phys Res Sect A Accel, Spectrometers, Detect Assoc Equip 479(1), 117–232. http://linkinghub.elsevier.com/retrieve/pii/S0168900201020137, https://doi.org/10.1016/S0168-9002(01)02013-7
3. Bevan AJ, Golob B, Mannel T, Prell S et al (2014) The physics of the B factories. Eur Phys J C 74(11), 3026. arXiv:1406.6311, https://doi.org/10.1140/epjc/s10052-014-3026-9
4. Abe T, Adachi I, Adamczyk K, Ahn S et al (2010) Belle II technical design report. arXiv:1011.0352
5. Wikipedia contributors (2018) KEKB (accelerator). https://en.wikipedia.org/wiki/KEKB_(accelerator)
6. SuperB Collaboration, Baszczyk M, Dorosz P, Kolodziej J et al (2013) SuperB technical design report. arXiv:1306.5655
7. Keck T, Abudinen F, Bernlochner FU, Cheaib R et al (2018) The full event interpretation—an exclusive tagging algorithm for the Belle II experiment. arXiv:1807.08680
8. Kemmer J, Lutz G (1987) New detector concepts. Nucl Instrum Methods Phys Res 253, 365–377. https://ac.els-cdn.com/0168900287905183/1-s2.0-0168900287905183-main.pdf?_tid=9c6da4c9-b91f-464b-ae11-89016a3e2531&acdnat=1524476602_881ecbe6d098bc1af5882f526389be5a
9. Nakao M, Higuchi T, Itoh R, Suzuki SY (2010) Data acquisition system for Belle II. J Instrum 5(12), C12004–C12004. https://doi.org/10.1088/1748-0221/5/12/C12004
10. Yamada S, Itoh R, Nakamura K, Nakao M et al (2015) Data acquisition system for the Belle II experiment. IEEE Trans Nucl Sci 62(3), 1175–1180. http://ieeexplore.ieee.org/document/7117478/, https://doi.org/10.1109/TNS.2015.2424717
11. Sun D, Liua Z, Zhao J, Xu H (2012) Belle2Link: a global data readout and transmission for Belle II experiment at KEK. Phys Procedia **37** (2012) 1933–1939, 2012. https://ac.els-cdn.com/S1875389212019104/1-s2.0-S1875389212019104-main.pdf?_tid=85c434c2-13bd-4420-a260-6c3fdce60e67&acdnat=1524477027_6fba611c71e0ba0d469f767f5c915f0c https://doi.org/10.1016/j.phpro.2012.01.036
12. Hearty C (2018) Jira issue on HLT phase 2 menu. https://agira.desy.de/browse/BII-3617?focusedCommentId=44831&page=com.atlassian.jira.plugin.system.issuetabpanels%3Acomment-tabpanel#comment-44831
13. Itoh R (2012) DQM and prompt/express reco. https://indico.phys.hawaii.edu/getFile.py/access?contribId=28&resId=0&materialId=slides&confId=290
14. The Belle II Computing Steering Group (2017) Belle II computing resources estimates for 2017–2021. https://docs.belle2.org/record/428/files/BELLE2-NOTE-TE-2016-012.pdf
15. Moll A (2011) The software framework of the Belle II experiment. J Phys Conf Ser 331(3), 032024. http://iopscience.iop.org/article/10.1088/1742-6596/331/3/032024/pdf, https://doi.org/10.1088/1742-6596/331/3/032024

16. Kuhr T, Pulvermacher C, Ritter M, Hauth T, Braun N (2018) The Belle II core software. arXiv:1809.04299
17. Standard C++ Foundation, The standard: standard C++. https://isocpp.org/std/the-standard
18. Python Software Foundation, Welcome to Python.org. https://www.python.org/
19. Brun R, Rademakers F (1997) ROOT–An object oriented data analysis framework. Nucl Instrum Methods Phys Res, Sect A Accel, Spectrometers, Detect Assoc Equip 389(1–2), 81–86. https://doi.org/10.1016/S0168-9002(97)00048-X
20. Guennebaud G, Benoît J (2010) Eigen v3. http://eigen.tuxfamily.org
21. Lange DJ (2001) The EvtGen particle decay simulation package. Nucl Instrum Methods Phys Res, Sect A Accel, Spectrometers, Detect Assoc Equip 462(1–2), 152–155. https://ac.els-cdn.com/S0168900201000894/1-s2.0-S0168900201000894-main.pdf?_tid=10489e10-9189-4f5f-acf7-87cedf64b28c&acdnat=1524480961_905eab864acd43a4f358afa21cf979c2, https://doi.org/10.1016/S0168-9002(01)00089-4
22. Dawes B, Abrahams D, Rivera R (2007) Boost C++ Libraries. https://www.boost.org/
23. ROOT Developers (2018) Cling | ROOT a data analysis framework. https://root.cern.ch/cling
24. Project Jupyter (2018) Jupyter homepage. http://jupyter.org/
25. Braun N, Hauth T, Pulvermacher C, Ritter M (2017) An interactive and comprehensive working environment for high-energy physics software with python and jupyter notebooks. J Phys Conf Ser 898(7), 072020. http://stacks.iop.org/1742-6596/898/i=7/a=072020?key=crossref.eff9ad3575e53bda6c70f5408e03c2cf, https://doi.org/10.1088/1742-6596/898/7/072020
26. Lee S, Itoh R, Katayama N, Mineo S (2011) Development of high level trigger software for Belle II at SuperKEKB. J Phys Conf Ser 331(2), 022015. http://stacks.iop.org/1742-6596/331/i=2/a=022015?key=crossref.6a11c19b6bd46d9524be163e2f3f12c6, https://doi.org/10.1088/1742-6596/331/2/022015
27. Urquijo P, Ferber T (2016) Overview of the Belle II physics generators. https://docs.belle2.org/record/282/files/BELLE2-NOTE-PH-2015-006-v2.pdf
28. Sjöstrand T, Mrenna S, Skands P (2007) A brief introduction to PYTHIA 8.1. arXiv:0710.3820, https://arxiv.org/pdf/0710.3820.pdf
29. Agostinelli S, Allison J, Amako K, Apostolakis J et al (2003) Geant4—a simulation toolkit. Nucl Instrum Methods Phys Res Sect A Accel, Spectrometers, Detect Assoc Equip 506(3), 250–303. https://ac.els-cdn.com/S0168900203013688/1-s2.0-S0168900203013688-main.pdf?_tid=2ab2887a-7a2b-4813-9966-25cc44845930&acdnat=1524482129_7273334d05ad6f2c66d093ae3c67b132, https://doi.org/10.1016/S0168-9002(03)01368-8
30. Eliachevitch M (2018) Tracking performance studies on cosmic rays and a track quality estimation for the central drift chamber of the Belle II-experiment. Master's thesis, KIT. https://ekp-invenio.physik.uni-karlsruhe.de/record/49043

Chapter 3
Foundations

The following chapter summarizes the basic mathematical and algorithmic foundations used throughout this thesis. The goal is to present the topics to a level of detail which helps to understand the work presented in the following chapters. For a more complete and thorough discussion, references on further reading are given in each section.

A large fraction of this thesis focuses on the development and validation of tracking algorithms (see Chap. 6). An introduction to tracking concepts is given in Sect. 3.1 and a more detailed introduction into the track finding algorithms for Belle II is given in Sect. 3.2. Afterwards, the principles behind track fitting—the extraction of the parameter of a trajectory out of multiple point measurements—and its implementation in the Belle II software framework are described. The chapter ends with a short introduction to multivariate methods and the estimation of statistical uncertainties used in this work.

3.1 Track Reconstruction

As described in Chap. 2, the task of the tracking subdetectors is to measure the trajectories of all charged final-state particles by recording the spatial position of energy depositions in the sensor region while a magnetic field in z direction is applied. Using reconstruction software, the spatial measurements (*hits*) of a traversing particle are combined into sets (*tracks*) to extract the kinematic properties of the particle. This step in the reconstruction is called *tracking* and is often divided into the task of partitioning the hits into tracks (*track finding*) and calculating the track parameters (*track fitting*).

N. Braun, *Combinatorial Kalman Filter and High Level Trigger Reconstruction for the Belle II Experiment*, Springer Theses,
https://doi.org/10.1007/978-3-030-24997-7_3

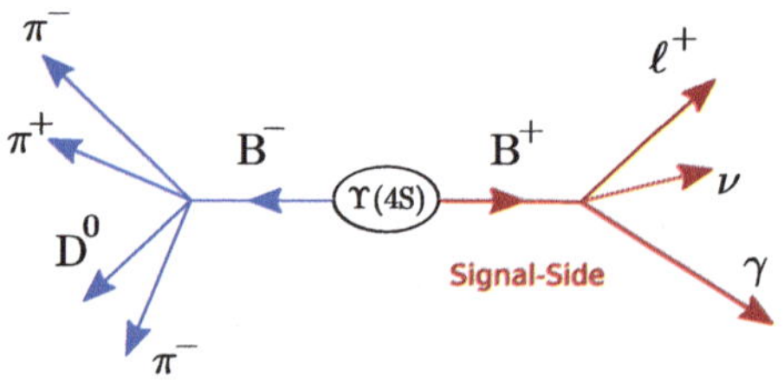

Fig. 3.1 Schematic decay of two B mesons into the very rare $B^+ \rightarrow \ell^+\nu\gamma$ final state on the one side and the more likely $B^- \rightarrow D^0\pi^-\pi^+\pi^-$ on the other side. Adapted from [1]

The tracking subdetectors in Belle II are the only devices which are able to reconstruct the momentum, charge and flight direction of all charged particles. Due to inefficient algorithms or detectors, not all tracks in the acceptance region of the tracking detectors can be reconstructed successfully. This behavior can be quantified with the track finding efficiency,[1] which plays a crucial role for every physics analysis which will be performed at Belle II. As physics analyses rely on the parameters of the produced particles, every efficiency in the analyses includes the track finding efficiency and the tracking systematics as a systematic uncertainty.

Apart from a high efficiency, also the quality of the track reconstruction results must fulfill certain requirements. To take advantage of the special feature of the experimental environment—namely the known initial state and the very clean decay of the $\Upsilon(4S)$ into exactly two B mesons—it is undesirable to produce additional tracks out of random combinations or to reconstruct the same particle multiple times. The influence of such mistakes during the tracking reconstruction can be seen when looking into a typical rare decay search like $B^+ \rightarrow \ell^+\nu\gamma$ (see Fig. 3.1), as it will be performed with the Belle II data [1]. An efficient background suppression is crucial for reconstructing the very rare decay. For this, the full event with both B mesons is reconstructed (see [2] for more information on the Full Event Interpretation). Signal events are required to have no remaining additional tracks in the event after the full event is interpreted. Mistakes in the tracking reconstruction as additional or missing tracks would deteriorate the full event interpretation and thus the signal reconstruction. Additionally, the resolution of the extracted properties of the particle such as the vertex position or the momentum are crucial for the physics analyses.

The reconstruction of measurements in the outer subdetectors, e.g. the TOP, relies on the extrapolation of the tracks into the volumes of these detectors. Imprecise track parameters can render it impossible to add the outer subdetector information to these particles. This is also true for the PXD reconstruction, which needs an already found track in the outer subdetectors for data acquisition as described in Sect. 2.1.

There are additional requirements with a more technical scope as tracking is part of the software for the online reconstruction and filtering. This requires a fast execution and a low memory footprint.

[1]More details on the definition of the finding efficiency can be found in Sect. 3.2.3 below.

3.2 Track Finding Concepts at Belle II

This section describes the algorithms used for track finding at Belle II, which are the input or the baseline for the algorithms presented in this thesis. The new algorithms described in the following chapters of this thesis are now part of the default event reconstruction and the old algorithms pictured in this chapter will be called *legacy* implementation. The described tracking algorithms are used for reconstructing MC generated and recorded events—differences in the online software are discussed in Chap. 4.

Belle II's track finding algorithms follow the separation imposed by the detector. The different characteristics and geometries of the subdetectors do not allow to find the tracks with one algorithm at the same time. The silicon-based subdetectors have a very good spatial resolution. Due to the geometry of the layers, a passing particle leaves typically only one cluster per sensor layer. This fact can be exploited during track finding. However, they are stronger affected by beam-induced background because of their proximity to the interaction point. Algorithms to handle the large number of hits and combinatorics are needed, which can however use very precise spatial information. The central drift chamber on the other hand has a very large number of layers and therefore a traversing particle produces many hits. On the downside, its single measurements do not have such a good spatial resolution compared to the measurements of the vertex subdetectors.

In the following, the three distinct track finding algorithms mentioned in Fig. 3.2 are described in more detail.

3.2.1 *Track Finding in the CDC*

The track finding using only information from the CDC subdetector is split up in two parts: a global algorithm using the Legendre transformation and a local one based on a Cellular Automaton. Both algorithms have a different sensitivity and cover different regions of the phase space. The Legendre algorithm applies a global mathematical transformation to the positions of all hits simultaneously. It is therefore fairly stable

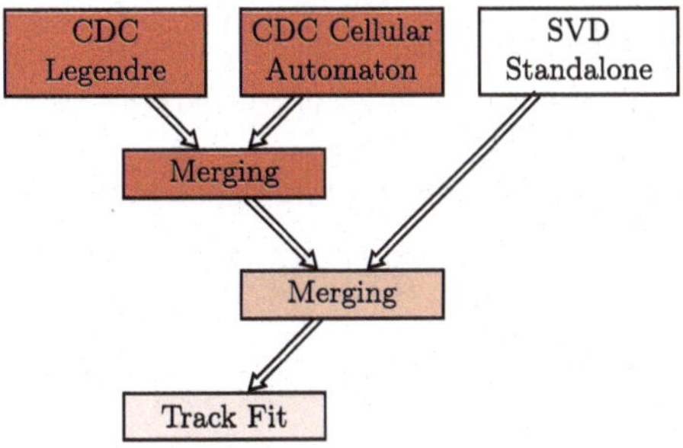

Fig. 3.2 Implemented tracking algorithms in basf2 and their interplay. The trajectories of the particles are reconstructed standalone in the different subdetectors and combined afterwards

in environments with high hit inefficiencies or when many particles cross each other in the same spot, which both lead to gaps in the tracks. On the other hand, this algorithm has a strong dependency on the modeled form of the trajectory which is assumed to be a circle through the IP without considering any energy loss or multiple scattering. The local algorithm uses local neighborhood relations and does neither depend on a specific form of the trajectory nor does it make any assumptions on the trajectory parameters. It is therefore a good complementary approach.

In basf2, both implemented algorithms operate on a prefiltered hit sample. Measured energy depositions likely produced by beam-induced background processes are removed using a multivariate method [3].

3.2.1.1 Global Legendre Algorithm

The global algorithm used for Belle II [4] is based on a paper by Alexopoulos et al. [5] and is an extended version of a transformation first described by Hough [6]. Neglecting interactions with the gas and the very sparse wires in the CDC, the path of a charged particle can be described as a circle in the r-ϕ plane. For non-secondary decays, this circle goes through the IP. Each drift circle produced by this traversing particle touches the trajectory of its track. An oversized example of four trajectories, which all touch a single drift circle and go through the IP is shown in Fig. 3.3. The Legendre transformation given by

$$\rho_{\pm}(\theta) = \frac{2}{x^2 + y^2 - d^2}\left(x\cos(\theta) + y\sin(\theta) \pm d\right) \qquad \theta \in [0, \pi],$$

transforms the position x, y and the measured drift length d of a single drift circle into the Legendre space of θ and ρ. Each point in this space describes a circle through the IP, where θ denotes the angle of the tangent to the circle in the origin and $\rho = 1/R$ is the (signed) curvature of the circle. The circles described by pairs $(\theta, \rho_{\pm})$ given by the transformation above touch the transformed drift circle. The two functions ρ_+ and ρ_- describe the orientation of the circle with respect to the hit. They are also depicted in Fig. 3.3.

If the trajectory parameters do not change along the flight path of a particle, it can be described by a single pair (θ, ρ) of circle parameters. If all hits of a track are transformed in the Legendre space, a search for the accumulation point leads to those parameters together with the information which hits make up this track. A bisection method in two dimensions is used for this. Some steps in this bisection method can be seen in Fig. 3.4.

The algorithm strongly depends on the fact that the trajectory is exactly circular and passed the interaction point. Different adjustments to the algorithm have to be made to account for deviating tracks. More information can be found in [4].

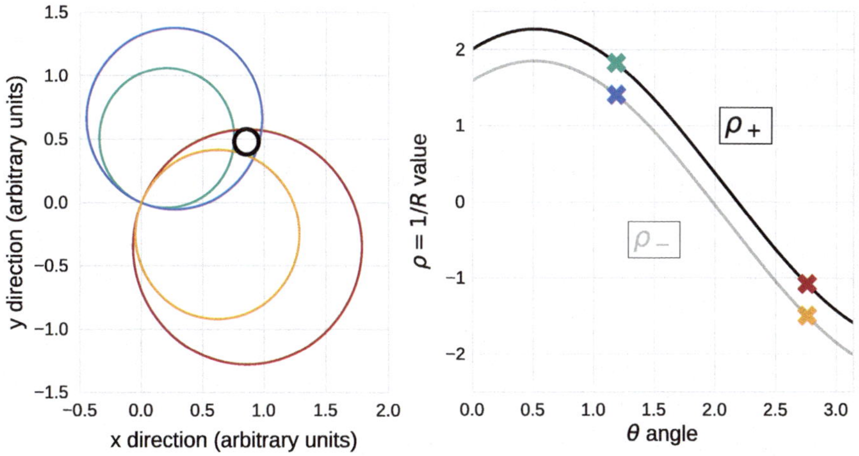

Fig. 3.3 Exemplary transformation of the position and drift length of a single (oversized) drift circle into the Legendre space of θ and ρ. The left figure shows the measured hit with four possible circles through the origin (in colors) touching the drift circle. The IP is chosen to be at $(0, 0)$. The right figure depicts the functions $\rho_{\pm}$ with the parameters of the four circles highlighted in the same colors

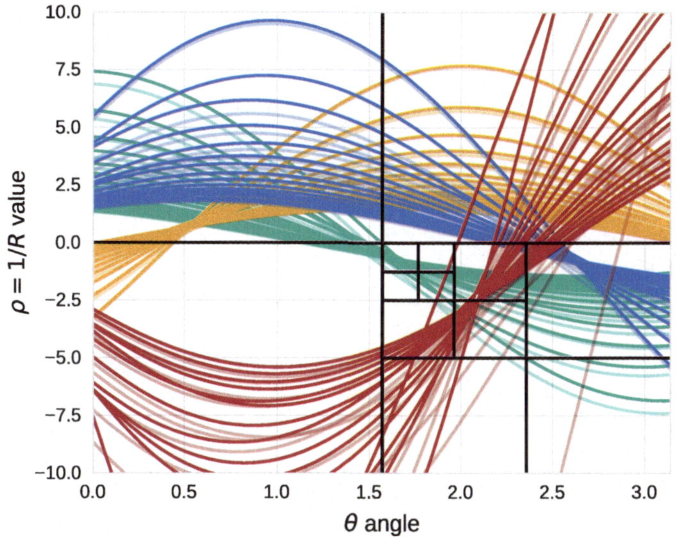

Fig. 3.4 Artificial example for four steps in the two-dimensional bisection method searching for the accumulation point of the green lines in the Legendre parameter space. The different colors depict (θ, ρ)-parameter sets from different tracks

As the presented algorithm relies on the x and y position of the measured drift circles, which are undefined for stereo hits, this procedure can only be applied to axial hits. It is however possible to add stereo hits by employing a similar method to already found axial tracks, which is further described in [3].

3.2.1.2 Local Cellular Automaton

To find tracks which deviate too strongly from the requirements imposed by the Legendre algorithm, a local algorithm is used, which is in further detail described in [7]. The method builds relations between neighboring hits and collects possible paths respecting those relations to form tracks. As seen in Fig. 3.5, the term neighboring is used here in a loose meaning, as also two hit wires are called neighboring if there is a passive wire between them. It is not possible to limit the relations between two spatial neighboring hits reasonably by only using two measurement points. Therefore, the two-hit relations are expanded to three-hit relations (*triplets*) and clusters spanning a full superlayer. During this process, cuts calculated on the features of the different hit combinations are employed, e.g. a common least squares fit.

Within each cluster, defined as the largest set of neighboring hits in a superlayer, the best path of hits must be found. This is done using the concept of a Cellular Automaton [8] known from computer science, which was already employed successfully to the track finding problem domain in other experiments (see [9, 10]).

The track finding problem is modeled as a graph, where each triplet of hits in the cluster is represented by a node. Each node has an edge to another node, if and only if the two related triplets share exactly two hits. The weights of this directed, acyclic graph[2] represent how well two triplets fit together, e.g. based on a common circular fit or even a multivariate method. The track finding task is then reduced to find the connected path over a set of nodes with the highest sum of weights in this graph. This is performed using the algorithm depicted in Fig. 3.6.

Starting with a randomly chosen start node, the path with the highest weight starting from this node is found recursively, by accessing the related child nodes in the graph. By passing the information on the best path to the parent nodes, the globally best path can be found. By removing the edges in this path and repeating the process, multiple results can be extracted. Those paths are called *segments*. After the segments are found, these can either be used to form new, yet unfound tracks or to improve the results of the Legendre algorithm. Both possibilities are used and leverage several multivariate methods for deciding on each segment, which is described in more detail in [3].

After both algorithms performed their finding procedure and the results are merged together, a final quality control process cuts away possible fake tracks and cleans up the tracks by removing wrongly assigned hits [3].

3.2.2 *Track Finding in the VXD*

The algorithm employed for SVD standalone track finding before the work of this thesis is called VXDTF2 [11]. It utilizes the same concept as the local Cellular

[2] A direction is imposed by the natural flight direction of the particles coming from the interaction point and the acyclicity can easily be ensured in the implementation without loosing efficiency.

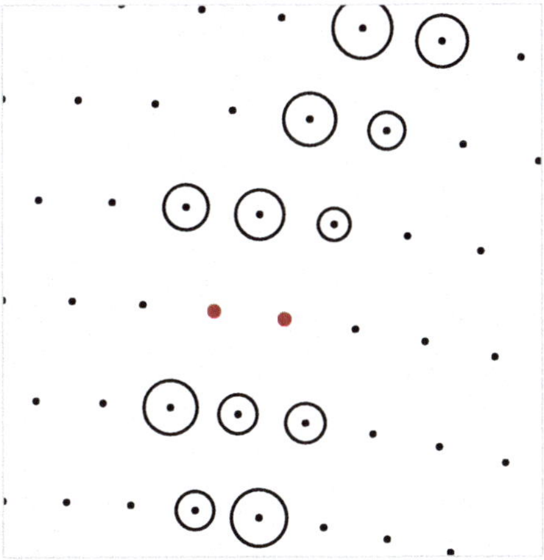

Fig. 3.5 A set of wires where signal particles deposited energy (shown as black circles) which are called neighboring although there are two missing hits in between (depicted as red wires)

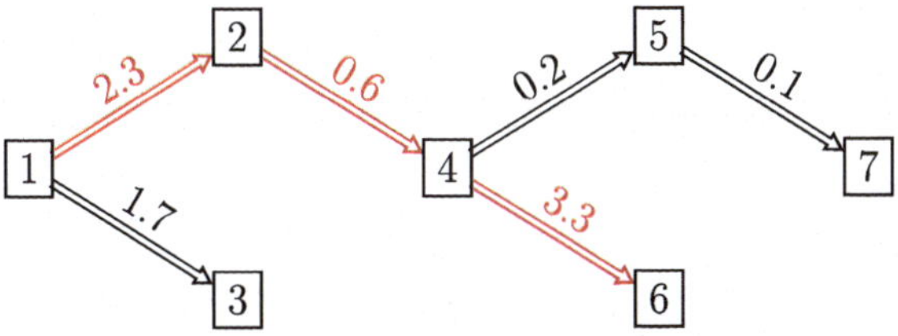

Fig. 3.6 Example for a graph used in the Cellular Automaton procedure. The boxes are the nodes and the lines show the edges with weights between them. The final result of the first iteration—the path with the highest sum of weights—is highlighted. After [3]

Automaton described above: Two- and three-hit combinations are built using filters to limit combinatorics. Because of the high occupancy of the SVD compared to the CDC, these filters are more complex and are described in the following. After the three-hit combinations are built, a Cellular Automaton is employed to extract possible track candidates as described above. All steps are shown schematically in Fig. 3.7.

The global Legendre algorithm described above for the CDC is unusable in this case, as the radius of the VXD is smaller than the CDC. The estimation of ρ needed in the Hough algorithm has therefore a very bad resolution, increasing the combinatorics far too much.

3.2.2.1 Filters

To utilize the Cellular Automaton approach, filters for relations among hits must be applied. A staged approach as in the CDC is used to gradually use the full detector

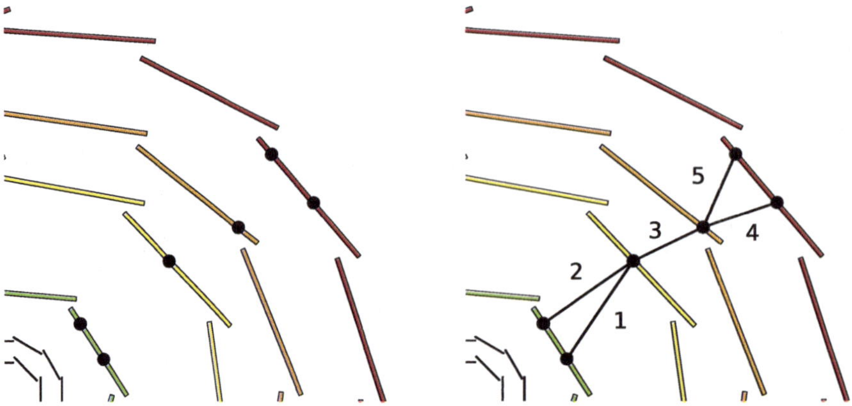

(a) Space point creation. (b) Two hit combinations.

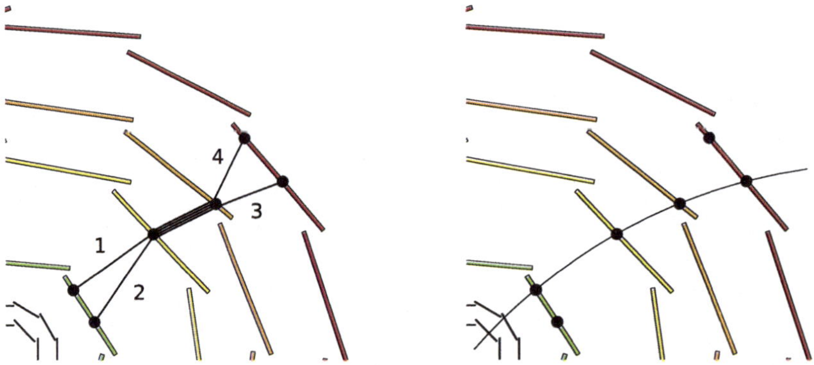

(c) Three hit combinations. (d) Final track.

Fig. 3.7 Basic steps in the VXDTF2. For better visibility, the steps are only shown for a few hits in the SVD. First (**a**), the two-dimensional strip measurements are combined into three-dimensional space points. Second, two- and then three-hit combinations (**b**, **c**) are built applying filters from the `SectorMap` at each stage. The final track (**d**) is found using a Cellular Automaton and an overlap reduction

information without increasing the combinatorics dramatically. The different stages are given in the following.

1. **Space points** All compatible u- and v- strip measurements in the SVD are combined to so-called *space points*. As the assignment between the two-dimensional strips is not unambiguous, wrong combinations creating ghost hits increase the number of space points to test.
2. **Two-hit filters** Using geometric constraints from the detector and variables from the recorded energy depositions, feasible combinations of two space points

are built. As the geometry of the detector is rather complicated, a special technique is applied (see below).

3. **Three-hit filters** Every pair of two-hit combinations sharing one hit and fulfilling another filter stage is then stored as a three-hit combination. Here again geometrical constraints as well as variables from the measurements are employed to decrease the combinatorics while keeping the efficiency high enough.

The two filter steps needed in the approach are trained using time consuming MC simulations of the full detector. As there are no simple symmetries to exploit in the detector, geometric cuts for the filters would be very complicated to implement. Additionally, the cuts on the measured features of the space points cannot be calculated from first principles as the full material effect handling is non-trivial making a simulation necessary anyway.

For this, the sensor planes of the SVD are divided into a grid of finer sectors. Friendship relations are defined between the sectors and stored in a so-called `SectorMap`. All hits from a sector pair combination, which was seen in the training set of MC particles, can form a valid two hit pair in the application phase. Together with information if two sectors are related, the values of the features for the two- or three-hit combinations in the MC simulation are stored, making it possible to perform cuts dependent on the location in the detector.

3.2.2.2 Cellular Automaton and Final Track Selection

Using the three hit combinations produced in the previous step, track candidates are created using a Cellular Automaton. In contrast to the way this is done in the CDC, not only the best candidate respecting the three-hit combinations is returned, but all candidates also including valid sub-paths are stored [12]. This allows to produce shorter track candidates where one hit may be missing because of high multiple scattering or detector inefficiencies.

Among the large set of candidates, a smaller final subset must be extracted. As the spatial resolution of the measurements is good, the probability of two distinct particles sharing more than a single space point is very low. This is why the final selection extracts the best set of candidates with at maximal one common hit between each pair of tracks. The quality of the track candidates is hereby evaluated using a least-squared fit or a multivariate method.

Although technically possible, the current anticipated level of beam-induced background in the PXD does not allow the VXDTF2 to run on the combined set of PXD and SVD hits with a reasonable efficiency and purity. It is therefore only used for SVD standalone track finding.

3.2.3 MC Matching and Performance Indicators

A crucial step in developing a tracking algorithm is the performance evaluation to ensure a high efficiency and purity for later physics analyses. Although also possible on recorded data, it is beneficial and necessary to measure the efficiency on simulation. The process of simulating events was already described in Sect. 2.3.2.

As a first step, a MC track is formed out of the true energy depositions of a single simulated particle and compared to tracks found by the algorithm—also called pattern recognition (PR) tracks. Only MC tracks with at least five degrees of freedom (NDF) are taken into account. Particles with a smaller number of degrees of freedom cannot be used in the track fit so finding them would not bring any benefit. This factors out the acceptance of the detector and makes a comparison well defined, as 100% efficiency is reachable. The number of shared hits N_s between a MC and a PR track is counted. By dividing this number with the total number of hits in the MC track, the *hit efficiency* of this MC-PR pair is built. Analogous, the *hit purity* is formed by dividing with the total number of hits in the PR track. Hereby, the NDFs for each hit are taken into account, e.g. a PXD hit is weighted twice as much[3] as a CDC hit. The tracks from both sets are then classified by these two numbers, as can be seen in Table 3.1. Averaging those classifications on a larger sample of events, different figures of merit can be calculated.

The *finding efficiency* describes the ratio between found and all MC tracks. It is the basic performance measure of the algorithm that goes into each later physics analyses and should be as high as possible. As the detector acceptance is already factored out (see above), a finding efficiency of 100% is theoretically reachable.

The *clone* and *fake rate* are the fractions of PR tracks classified in this category as shown in Table 3.1.[4] As described at the beginning of this chapter, many analyses depend on a low number or even zero wrong tracks in the event, so the fake and clone rate should be as low as possible.

The *hit efficiency* is the averaged single track efficiency (as mentioned in the top row of Table 3.1) for all found MC tracks in the sample. It can also be calculated separately for the different subdetectors. It is worth mentioning that this number is not the ratio of found hits in the event, as it is only calculated on the MC tracks classified as found. This means an algorithm with a lower finding efficiency may even have a better hit efficiency, although in total less correct hits are collected. An increasing number of correct hits per track improves the resolution (see below) and other derived quantities like the particle identification, which then in turn have a large impact on physics analyses.

Similar to the hit efficiency, the *hit purity* is the averaged single track purity for all matched or clone PR tracks. A low hit purity has a large impact on the resolution, although it strongly depends on the location in the detector where the wrong hits are

[3] As the PXD plane is given by the detector layout, a single PXD hit gives a two-dimensional spatial information. A CDC hit with a measured drift length only leads to a one-dimensional measurement.

[4] Sometimes the clone rate is given with respect to all matched or clone tracks instead of all tracks disregarding fakes. This convention is not used in this thesis.

Table 3.1 Different classification of MC (left side) and PR (right side) tracks based on the efficiencies and purities calculated between them as indicated by the arrows. The meaning is described in the first row of the table. Adapted from [3]

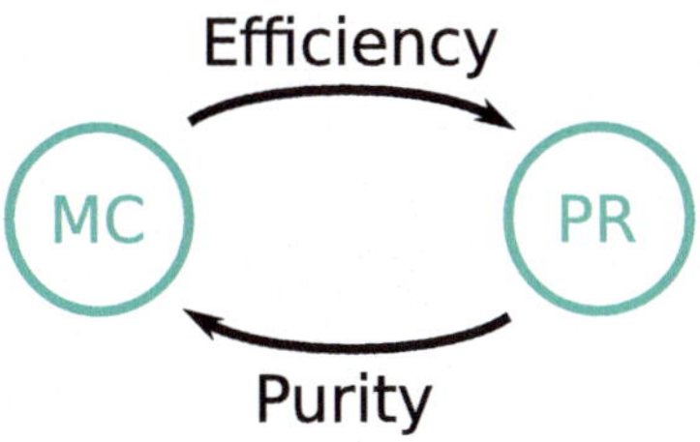	Example for the assignment between the truth information (represented by the MC tracks) and the tracks found by the tracking algorithm (PR tracks). The efficiency/purity is the number of shared hits divided by the total number of hits in the MC/PR track
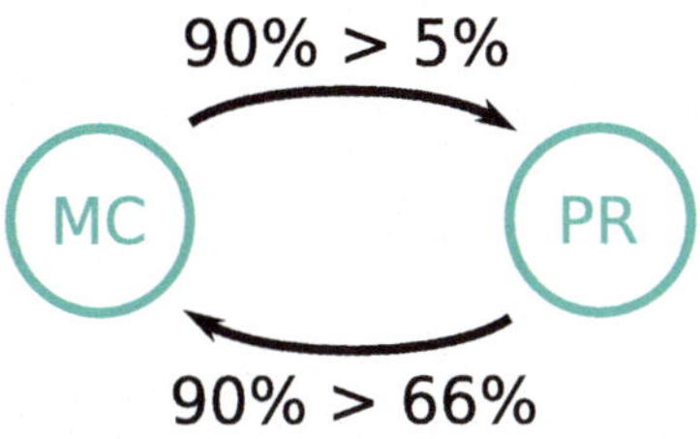	If a MC track is related to at least one PR track, with a minimum purity of 66% and a minimum efficiency of 5%, it is called found. A PR track fulfilling these criteria is called matched. There is a certain arbitrariness in the chosen minimal requirements
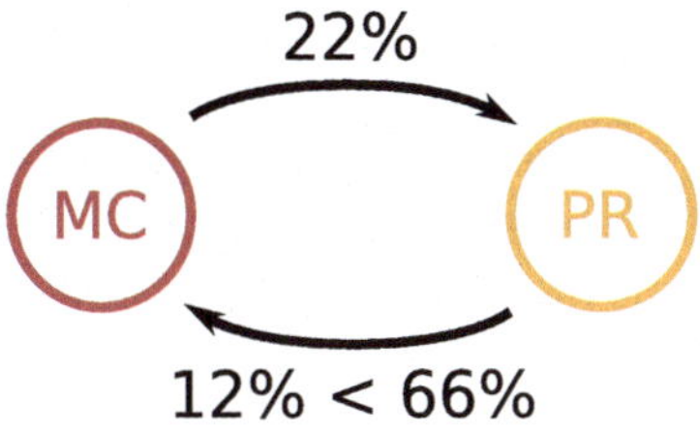	A PR track can fall below the purity requirement, if too many hits from beam-induced background processes are attached to it. It is then called background or fake track. If there is no matched track for a MC track, it is classified as missing
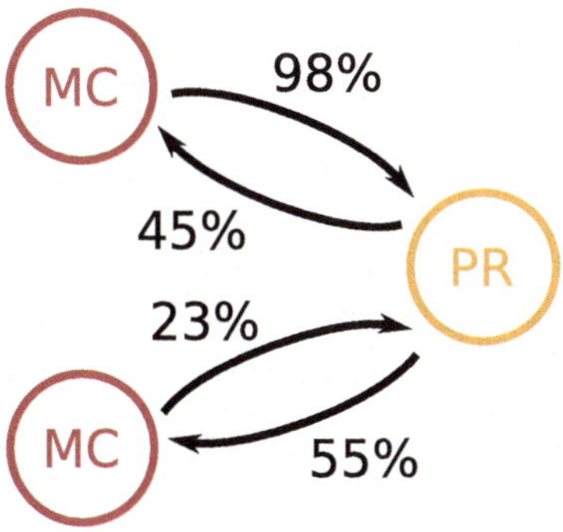	The PR track may also be a fake track, if it includes hits of multiple MC tracks and none is above the purity and efficiency threshold. This may happen when a decay in flight took place, where the child particle has more or less the same trajectory as its parent. As this case can be quite frequent, this behavior can also be turned off—merging the two MC tracks into one single track

(continued)

Table 3.1 (continued)

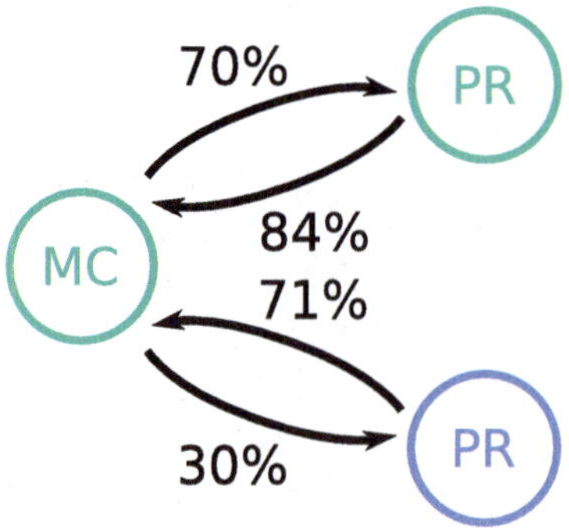	If in turn the MC track was found by more than one PR track, one is called found, whereas all other PR tracks with less efficiency or purity are classified as clone. This is a very typical case for low-momentum tracks, which curl in the tracking volume many times or if the MC particle is reconstructed separately in two subdetectors without merging the two pieces together

picked up. The averaged hit purity can never be lower than 66%, as below a track would be classified as a fake.

When a correct relation between a PR and a MC track is made (which is only true for matched PR tracks), the trajectory parameters reconstructed by the algorithm can be compared to the truth parameters used in the simulation of the particle. A good resolution on these parameters is crucial for physics analyses as many derived quantities, e.g. the invariant mass, depend on the tracking results. Also, vertexing may not be applicable with imprecise parameter estimations. The resolution is defined as the deviation between the measured and the truth value of a certain quantity. Most interesting in the context of this thesis is the resolution of the helix parameters (see Sect. 3.3.1). Sometimes, the pull distributions is shown, which is build by normalizing the resolution distribution to the uncertainty of the measured quantity. For Gaussian uncertainties, the pull distribution follows a standard normal distribution.

3.3 Track Fitting

After the hits in the detector are grouped into sets of tracks, this information is used to extract the parameters of the trajectory—the track fitting process. During this process, a correct handling of measurement uncertainties and all physical effects disturbing the helical path of the track have to be taken into account to arrive at a correct description of the trajectory. Those effects include interactions with the material such as energy loss, multiple scattering which depends highly on the geometry of the detector and the inhomogeneous magnetic field. Taking all these effects into account renders the mathematical description of the track model very complicated and allows only a numerical calculation. Dedicated fitting algorithms for this task were developed or adjusted, as described in this section.

3.3.1 Tracking Parameters

The primary task of tracking is to extract the momentum as well as the position of the particles. As the trajectory of charged particles in a homogeneous magnetic field can be modeled in first order as a helix, a description following the typical helix parameters $d_0, z_0, \tan\ \lambda, \phi_0, \omega$ is appropriate. The five quantities described in the following and depicted in Fig. 3.8 are used throughout this thesis. The same conventions as in the former BaBar experiment [13] are used. As the real trajectory is not a perfect helix, these five quantities change along the flight path of the track. If a concise description is needed, the helix parameters at the point of closest approach (POCA) to the IP are given, as those describe the produced particle in the collision best. For the definition of the coordinate system see Sect. 2.1.2.

$\boldsymbol{d_0}$ The signed distance of the POCA to the IP in the x–y plane, which is by definition the minimal distance. The sign corresponds to the sign of the angular momentum at this point. The real trajectory of the particle does not need to pass this point, which may happen for secondary particles, where the production vertex is further away from the IP.

z_0 The distance in the z direction of the POCA.

tan λ The angle λ is defined between the x–y plane and the tangent to the helix at the POCA, which describes the deflection in z direction.[5] λ takes values between 0 and π. Because of the uniform motion of the particles in z direction, $\tan\lambda$ is equal to $\mathrm{d}z/\mathrm{d}s$, where s is the arc length.

ϕ_0 The angle between the tangent to the helix at the POCA in the x–y plane and the x axis (by convention). It is defined between 0 and 2π.

ω The signed curvature of the helix in the x–y plane. The signature is defined by the sense of rotation, which is the charge of the particle.

All other properties, e.g. the transversal momentum, can be derived from those five basic parameters. These calculations are not repeated here and can be found elsewhere in great detail (e.g. [14]).

3.3.2 Track Fit Using χ^2

A typical procedure to extract parameters of a model from measured data is to find the parameters that minimize the squared distance between the recorded data points and the evaluated model points shown in Fig. 3.9a. This distance is also known as χ^2 (see e.g. [15] or [16] for more information) and is defined as

$$\chi^2 = (\mathbf{y} - \mathbf{f}(\mathbf{x}))^T\, \mathbf{V}^{-1}\, (\mathbf{y} - \mathbf{f}(\mathbf{x}))\ ,$$

[5] Sometimes, the angle $\theta = \pi - \lambda$ is used instead.

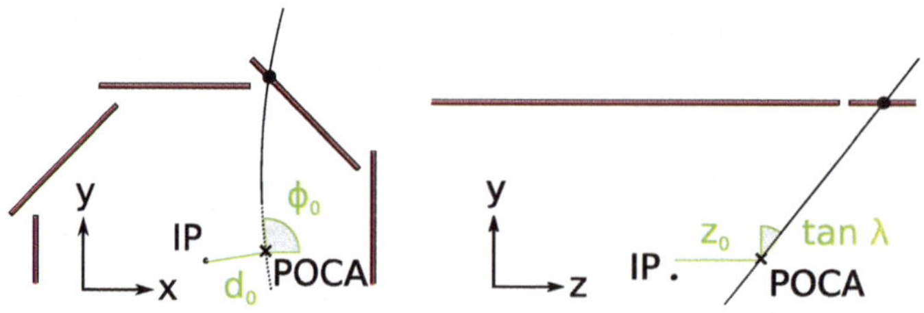

(a) View in the x–y plane. (b) View in the z–y plane.

Fig. 3.8 Description of the helix parameters with the conventions used in this thesis and in Belle II. The fifth parameter ω is not shown in the figure. The first PXD layer is displayed for visual guidance

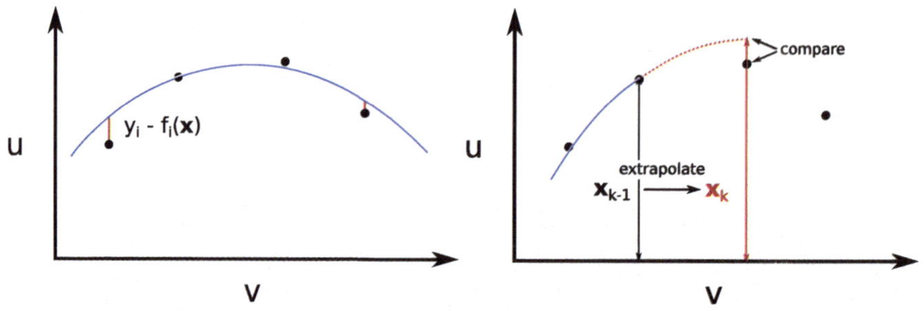

(a) χ^2 based method (b) Kalman filter based method

Fig. 3.9 Conceptual sketch of the χ^2 fitting method in comparison with the Kalman filter. In this simplified example, the χ^2 based method would be the computationally fastest choice, as the analytic form of the fit model is known

where $\mathbf{f} = (f_1, f_2, \ldots, f_n)$ denotes the functions to model the recorded data $\mathbf{y} = (y_1, y_2, \ldots y_n)$. The model $\mathbf{f}$ depends on m parameters $\mathbf{x}$, which need to be estimated. In most of the applications the f_i are given by the same function. The matrix $\mathbf{V}$ denotes the covariance matrix of the measurements and its off-diagonal entries are non-zero in cases of correlations between measurements. If more than one quantity is measured at a time—e.g. a two-dimensional spatial measurement—the measurements y_l and the matrix $\mathbf{V}$ are extended accordingly.

In cases where the function $\mathbf{f}$ is known analytically, the minimizing parameters can be calculated directly. When an approximate solution $\mathbf{x}_0$ is already known, a linearization around these parameters $\mathbf{x}_0 + \delta\mathbf{x}$ can be performed making the result also directly computable. In all other cases, an iterative approach must be taken, e.g. by using a gradient-descent method to minimize the χ^2 function.

The χ^2 method is not usable for the final track fit. First, there is no closed form for the track model, so an analytical solution is not usable. Second, because of the high deviations from the helical path caused by energy losses, the linearization is not a good approximation. Also, the calculation of this result includes the inversion of a matrix, which grows with the number of measurements and parameters, which can be quite large for typical tracks.[6] Last, the final remaining approach—iterating until a minimization of the χ^2 is found—is not feasible, as it includes the calculation of all f_l functions multiple times, which would result in a large computing time.

Although the χ^2 method has the described drawbacks for the final track fit, it is still used for intermediate fast fits without taking material effects into account [7]. Those results are not only important during track finding, where computing speed plays a crucial role as many of those fits need to be performed, but can also be used as input to other, more precise algorithms.

3.3.3 Kalman Filter

A possible solution to the mentioned problems was described by [17], which adapted the Kalman filter algorithm [18] for tracking problems. The Kalman filter, shown in Fig. 3.9b, uses an iterative approach, which only needs to calculate one f_l at the time by using only the information from the step before. This reduces costly calculations and also the size of the matrices, which need to be handled and inverted at each step. It is widely used for all kind of optimization problems in and outside of the physics domain.

The algorithm consists of several steps, which are iterated from the first to the last data point to be fitted and are described in Fig. 3.10. Before going into detail, several definitions are necessary. In most cases, the notation follows the original work in [17]. For comparison, those will be explained on the educational example of fitting a linear model onto measurements m_i of the y position at fixed discrete x values.[7]

At each iteration step i, the current set of parameters describing the state of the model is $\mathbf{x}_i$. In the example of a linear function, a natural choice would be the y position and the slope s,

$$\mathbf{x}_i = \left(y_i, s_i\right)^T .$$

[6] A typical track with 100 measurements of a two parameters each needs to invert a 200×200 matrix. The Kalman approach discussed below only needs to invert 2×2 matrices, for which analytical forms exist.

[7] The Kalman filter can also be introduced as the optimal estimator for a discrete linear dynamic system. More information on this can be found in many textbooks, e.g. in [19].

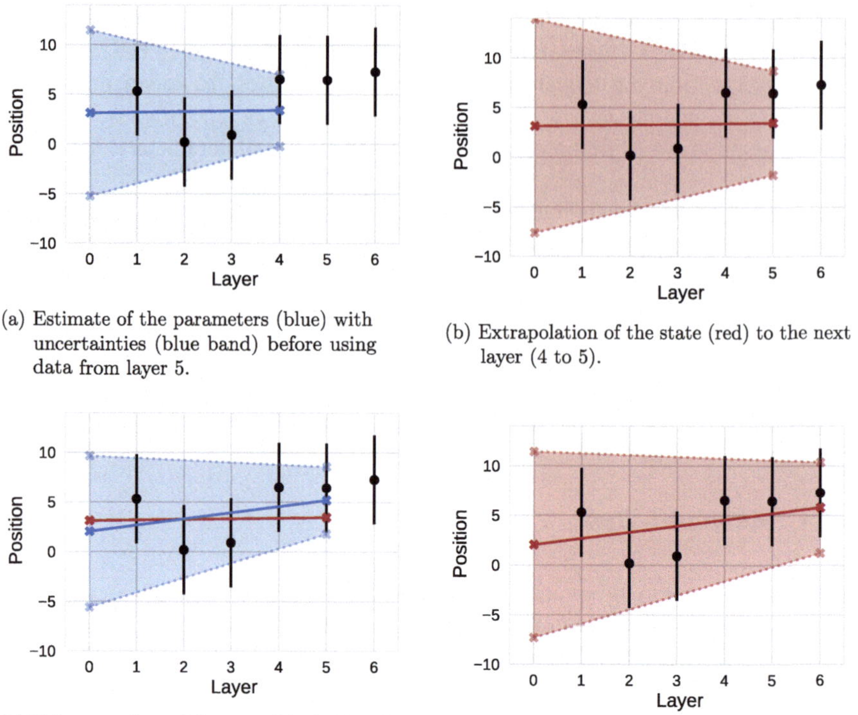

(a) Estimate of the parameters (blue) with uncertainties (blue band) before using data from layer 5.

(b) Extrapolation of the state (red) to the next layer (4 to 5).

(c) Kalman update of the state (blue) using the extrapolated position (red) and the measurement (black) on layer 5.

(d) Usage of the updated states for the next extrapolation (red).

Fig. 3.10 Example of one iteration in the Kalman filter procedure using toy data (in black). The shown positions and layers should not resemble actual detector data. A linear model is used for fitting the data

The state $\mathbf{x}_i$ changes when going from i to the next iteration $i + 1$, e.g.

$$\mathbf{x}_i \rightarrow \mathbf{x}_{i+1} = \begin{pmatrix} y_i + s_i \Delta x \\ s_i \end{pmatrix} = \mathbf{F}_i(\mathbf{x}_i) = \begin{pmatrix} 1 & \Delta x \\ 0 & 1 \end{pmatrix} \cdot \mathbf{x}_i$$

for the linear function example. This transformation from i to $i + 1$ is called *extrapolation*. Neither the step i, the x value or the distance between x values, Δx, are part of the state but are given by the description of the model and the measurements itself. The function $\mathbf{F}_i$ is dependent on the step i and given by the fit model. It can take on very complex forms, e.g. in the case of track fitting, it can only be calculated numerically. The uncertainty of the state $\mathbf{x}_i$ is given by the covariance matrix $\mathbf{C}_i$ and is also changing under extrapolation.

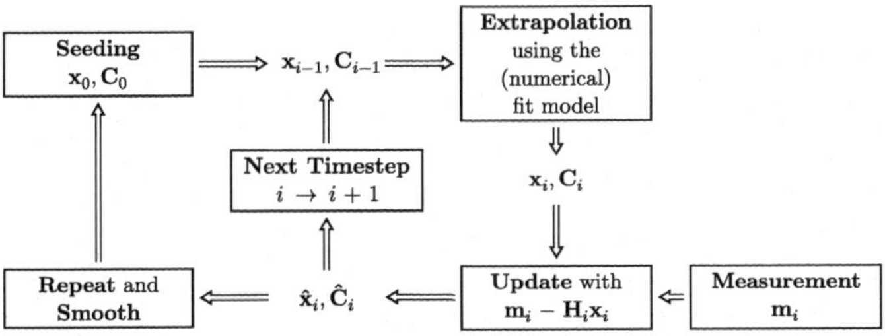

Fig. 3.11 Schematic Kalman filter procedure. Figure after [20]

The measurement performed at step i is $\mathbf{m}_i$ with the covariance matrix $\mathbf{V}_i$. In a general form, the measurements can be more dimensional, when measurements of multiple quantities are performed at the same time. For the linear function example, only the y position is measured, so the vector $\mathbf{m}_i$ is one dimensional.

Although in this example the measurement is a quantity of the state itself, in principle there can be a linear relation $\mathbf{H}_i$ between $\mathbf{m}_i$ and $\mathbf{x}_i$, when the state is not measured directly. In this simple example case, it is given by

$$\mathbf{H}_i = \begin{pmatrix} 1 & 0 \\ 0 & 0 \end{pmatrix}.$$

With these definitions, the Kalman filter procedure goes as follows (also described in Fig. 3.11):

Seed Finding At the first step of the Kalman filter, a state vector $\mathbf{x}_0$ which is used to extrapolate to the first measurement is needed. This first seed state does only need to be a rough estimate—so it is sufficient to use the tracking parameters extracted from a χ^2 based method without material effects taken into account. Typically, to minimize the dependence on the seed, the uncertainties of the state are artificially increased.

Extrapolation Using the fit model at each step, the state vector $\mathbf{x}_{i-1}$ and its covariance $\mathbf{C}_{i-1}$ are extrapolated to the step i.

Update After the extrapolation, the new state $\mathbf{x}_i$ is compared with the measurement $\mathbf{m}_i$ and updated to decrease the difference to the measurement. This is done using both the uncertainties of the measurement and the state as a weight $\mathbf{K}_i$

$$\hat{\mathbf{x}}_i = \mathbf{x}_i + \mathbf{K}_i\left(\mathbf{m}_i - \mathbf{H}_i\mathbf{x}_i\right) \qquad \hat{\mathbf{C}}_i = \left(\mathbb{1} - \mathbf{K}_i\mathbf{H}_i\right)\mathbf{C}_i.$$

The term $\mathbf{m}_i - \mathbf{H}_i\mathbf{x}_i$ is also referred as the residual $\mathbf{r}_i$. The weight $\mathbf{K}_i$ is also called the Kalman gain matrix and is given by

$$\mathbf{K}_i = \mathbf{C}_i\mathbf{H}_i^T\left(\mathbf{V}_i + \mathbf{H}_i\mathbf{C}_i\mathbf{H}_i^T\right)^{-1} = \mathbf{C}_i\mathbf{H}_i^T\mathbf{R}_i^{-1},$$

which is the covariance matrix of the state scaled with the one from the residual $\mathbf{R}_i$. The state is now a weighted average of the information of the current measurement $\mathbf{m}_i$ and the former measurements via the state $\mathbf{x}_i$. The updated state $\hat{\mathbf{x}}_i$ can then be used in the next extrapolation step until all measurements are consumed. The last state includes then information from all measurements and leads to the best estimation of the correct parameters describing the data.

Smoothing An even better estimation can be calculated by repeating the described iteration over all measurements multiple times. The result of a former run of the algorithm is used as the start seed for the next one. Again, the uncertainties are increased artificially to avoid an overestimation of the calculated error on the estimated parameters. As the extrapolation stopped with the last measurement, it is computationally beneficial, to repeat the algorithm in a reversed order. Doing so, one can use both passes in the two directions to built an average state at each measurement, which is called *smoothing*. This step also reduces the dependence of the important track parameters near the interaction point on the seed, which is also defined at this point.

The presented Kalman based fit method can directly be used for track finding and fitting. The different steps correspond to the different detector layers, where a spatial measurement of the track happened. The function used in extrapolation is given by the track model and is described further down.

3.3.4 Deterministic Annealing Filter

As described above, the Kalman filter has many advantageous attributes for using it in the track parameter estimation. The downside of the hit based methods is, that in principle every hit attached to the track influences the final track parameters. In consequence wrongly attached hits, e.g. hits coming from beam-induced background, can degrade the fit quality. Although there is a certain weighting in the Kalman filter, it is only based on the uncertainties of the measurement which are the same no matter if the hit is correctly assigned or not.

A solution to this problem is the deterministic annealing filter (DAF) [21]. First, it uses the Kalman filter based approach to calculate tracking parameters using all hits. With the first estimation, the residual between every hit and the fitted track is calculated and hits with a large residual (distance) r_{max} to the track are downweighted in the next iteration of the Kalman filter.

Because the first iteration in the Kalman filter is based on a seed state, which is only a rough estimate of the real parameters, it is necessary to change the definition of r_{max} in the different iterations. For this, an annealing scheme is applied. The maximal residual is dependent on an artificial temperature, which is decreased in each iteration leading to a more strict requirement on the hit residual. In the end, each hit is unambiguously attached to the track or not.

The algorithms described in this section are implemented into the basf2 framework using the genfit2 [22] library.

3.3.5 Extrapolation

As described above, a basic ingredient for the Kalman filter is the extrapolation of the state to the plane of the next subdetector including the measurement. The reason for choosing a plane is the state, which is used to describe the track parameters. It is given by

$$\mathbf{x} = \left(p/q, u', v', u, v\right)$$

and only defined on a plane perpendicular to the trajectory of the track. Choosing this state representation has several benefits, including simpler covariance calculations. The measurement plane is often given by the detector itself—e.g. in case of the sensor planes of the VXD of Belle II—or can be calculated using the current parameters of the track and the hit position—e.g. for the drift chamber measurements in the CDC. The problem of extrapolation is then reduced to transporting the state with its covariance from one of these planes to another, including material effects and the equation of motion in an inhomogeneous magnetic field.

The basic procedure to solve the equation of motion numerically, which is implemented in the genfit library, is the Runge-Kutta-Nyström method [23]. The method has access to the detector representation in the software to calculate material effects. Various material effects are taken into account such as the energy loss according to the Bethe formula, multiple scattering and soft bremsstrahlung energy loss [22]. Field inhomogeneities are included via the measured or calculated magnetic field map.

3.4 Mathematical Foundations

The last section of this chapter briefly introduces the concept of multivariate analysis as detailed as needed for following this thesis. There exists a lot of literature describing multivariate methods [16, 24], so only a short overview is given. This section also includes a description how statistical uncertainties for the MC studies are treated in this work.

3.4.1 Multivariate Analysis (MVA)

The decisions made by the tracking algorithm during the reconstruction depend on many input variables or *features*. Examples for these decisions include track-hit attachment, hit filtering and track quality estimation. Most choices are binary classifications, where the algorithm needs to decide to which class y an element described by the feature vector $\mathbf{x}$ belongs. In many cases, the two classes refer to a positive or a negative outcome—so they are called *signal* ($y = 1$) or *background* ($y = 0$).

Typically, the decisions are based on properly hand-crafted cuts on variables calculated either with physical knowledge or—as this is often impossible—using results of simulations. Often, the cuts do not take into account correlations between the variables. Also the dependence of the efficiency and the purity of the applied cut is unknown and it is hard to later tune the cuts, e.g. to achieve a higher efficiency in spite of lower purity.

A commonly used solution to such problems is the application of multivariate methods (see e.g. [16] for a detailed description), which approximate the decision function $f(\mathbf{x}) \to q$ that maps the vector of calculated features $\mathbf{x}$ to a test statistic q. The value q is the probability that the given element described by $\mathbf{x}$ belongs to the signal class $y = 1$. Performing a cut on q leads to a binary classifier, which is tunable for higher efficiency (lower cut values) or higher purity (higher cut values). The decision function f is fitted in the so-called training step using labeled data. The approximated function can then be used to predict the signal probability for unknown data points.

3.4.1.1 Boosted Decision Trees (BDTs)

The multivariate method employed in this thesis is taken from the FastBDT package [25], which implements a stochastic gradient-boosted decision tree [26]. A decision tree is a weak classifier, which partitions the feature space into hypercubes maximizing the ratio of a single class against the remaining one in each cube. The divisions for this partition are learned during the training phase by maximizing, e.g. the separation gain of each feature. Only shallow decision trees are utilized forcing the model to generalize. To retrieve a high decision power while being robust against overfitting, many decision trees are constructed where each tree is trained on a sample weighted by the wrong classifications of the previous one—a process known as boosting. The averaged result of all decision trees is a well-regularized and strong classifier.

3.4.2 *Treatment of Statistical Uncertainties*

Many results of this thesis rely on studies performed with MC simulation (see Sect. 2.3.2). As the samples used for those studies are finite, a statistical uncertainty on all calculated quantities emerges.

Basically, there are two different types of figures of merit, that will be discussed in this work. Because of their different underlying distributions, they have to be treated differently.

As described above, the finding efficiency as well as the clone and fake rate are calculated by dividing a number of tracks k in a certain category by the total number of tracks N, $\varepsilon = k/N$. There are a number of ways to calculate the uncertainty on ε, see for example [12]. The situation for hit efficiencies and purities is somewhat special. Those figures of merit are calculated by averaging the single track quantities over a larger set of tracks. However, the distribution of the single quantities is not known and can also not be calculated from first principles. Therefore, a method known as bootstrapping (see e.g. [27]) is used for all uncertainties in this thesis, which works also if the distribution is unknown. For calculating a feature $x(d)$ in a set of data points d_i, 1000 smaller subsets of the full data sample are built using random draws with replacement. On each of these subsets, the feature x is calculated giving a distribution of the feature instead of just a single value for the full dataset. The uncertainty is then given by the standard deviation of this distribution, which is approximately Gaussian distributed because of the law of large numbers.

References

1. Gelb MJ (2018) Search for the rare decay B+ to l+ nu gamma with the full event interpretation at the Belle experiment. PhD thesis, KIT. https://ekp-invenio.physik.uni-karlsruhe.de/record/49050/files/EKP-2018-00029.pdf, https://doi.org/10.5445/IR/1000085417
2. Keck T, Abudinen F, Bernlochner FU, Cheaib R et al (Jul 2018) The full event interpretation—an exclusive tagging algorithm for the Belle II experiment. arXiv:1807.08680
3. Braun N (2015) Momentum estimation of slow pions and improvements on the track finding in the central drift chamber for the Belle II experiment. Master's thesis, KIT. https://ekp-invenio.physik.uni-karlsruhe.de/record/48740/files/EKP-2015-00052.pdf
4. Trusov V (2016) Development of pattern recognition algorithms for the central drift chamber of the Belle II detector. PhD thesis, KIT. https://ekp-invenio.physik.uni-karlsruhe.de/record/48882/files/EKP-2017-00009.pdf
5. Alexopoulos T, Bachtis M, Gazis E, Tsipolitis G (2008) Implementation of the legendre transform for track segment reconstruction in drift tube chambers. Nucl Instrum Methods Phys Res Sect Accel Spectrometers Detect Assoc Equip 592(3):456–462. https://www.sciencedirect.com/science/article/pii/S0168900208005780, https://doi.org/10.1016/j.nima.2008.04.038
6. Hough PVC (1962) Method and means for recognizing complex patterns. Patent. http://www.google.com/patents/US3069654

7. Frost O (2013) A local tracking algorithm for the central drift chamber of Belle II. Master's thesis, KIT. http://ekp-invenio.physik.uni-karlsruhe.de/record/48172/files/iekp-ka2013-6.pdf
8. Toffoli T, Margolus N (1987) Cellular automata machines: a new environment for modeling. MIT Press. https://dl.acm.org/citation.cfm?id=22919
9. Abt I, Emeliyanov D, Kisel I, Masciocchi S (2002) CATS: a cellular automaton for tracking in silicon for the HERA-B vertex detector. Nucl Instrum Methods Phys Res Sect A Accel Spectrometers Detect Assoc Equip 489(1–3):389–405. http://linkinghub.elsevier.com/retrieve/pii/S0168900202007908, https://doi.org/10.1016/S0168-9002(02)00790-8
10. Kovalenko V (1997) Cellular automaton and elastic net for event reconstruction in the NEMO-2 experiment. Nucl Instrum Methods Phys Res Sect A Accel Spectrometers Detect Assoc Equip 389(1–2):169–172. http://linkinghub.elsevier.com/retrieve/pii/S0168900297000806, https://doi.org/10.1016/S0168-9002(97)00080-6
11. Lettenbichler J (2016) Real-time pattern recognition in the central tracking detector of the Belle II experiment. PhD thesis, Technische Universität Wien. http://www.ub.tuwien.ac.at
12. Wagner J (2017) Track finding with the silicon strip detector of the Belle II experiment. Master's thesis, KIT. https://ekp-invenio.physik.uni-karlsruhe.de/record/48930/files/EKP-2017-00057.pdf
13. Brown DN, Charles EA, Roberts DA (2000) The BaBar track fitting algorithm. In: Proceedings of computing in high energy physics conference, Padova
14. Avery P (1992) Applied fitting theory: IV formulas for track fitting. https://www.phys.ufl.edu/~avery/fitting/fitting4.pdf
15. Sorenson HW (1970) Least-squares estimation: from Gauss to Kalman. IEEE Spectr 7(7):63–68. http://ieeexplore.ieee.org/document/5213471/, https://doi.org/10.1109/MSPEC.1970.5213471
16. Blobel V, Lohrmann E (1998) Statistische und numerische Methoden der Datenanalyse. Teubner Studienbücher Physik. Vieweg+Teubner Verlag, Wiesbaden. http://link.springer.com/10.1007/978-3-663-05690-4, https://doi.org/10.1007/978-3-663-05690-4
17. Frühwirth R (1987) Application of Kalman filtering to track and vertex fitting. Nucl Instrum Methods Phys Res Sect A Accel Spectrometers Detect Assoc Equip 262(2–3):444–450. http://linkinghub.elsevier.com/retrieve/pii/0168900287908874, https://doi.org/10.1016/0168-9002(87)90887-4
18. Kalman RE (1960) A new approach to linear filtering and prediction problems. J Basic Eng 82(1):11. arXiv:NIHMS150003, http://fluidsengineering.asmedigitalcollection.asme.org/article.aspx?articleid=1430402, https://doi.org/10.1115/1.3662552
19. Gelb A (1974) Applied optimal estimation. MIT Press
20. Aimonen P (2011) Basic concept of Kalman filtering. https://commons.wikimedia.org/wiki/File:Basic_concept_of_Kalman_filtering.svg
21. Frühwirth R, Strandlie A (1999) Track fitting with ambiguities and noise: a study of elastic tracking and nonlinear filters. Comput Phys Commun 120(2–3):197–214. http://linkinghub.elsevier.com/retrieve/pii/S0010465599002313, https://doi.org/10.1016/S0010-4655(99)00231-3
22. Rauch J, Schlüter T (2015) GENFIT—a generic track-fitting toolkit. J Phys Conf Ser 608:012042. arXiv:1410.3698, http://arxiv.org/abs/1410.3698, https://doi.org/10.1088/1742-6596/608/1/012042
23. Lund E, Bugge L, Gavrilenko I, Strandlie A (2009) Track parameter propagation through the application of a new adaptive Runge-Kutta-Nyström method in the ATLAS experiment. J Instrum 4(04):P04001. http://stacks.iop.org/1748-0221/4/i=04/a=P04001?key=crossref.4db573cdf47b35d9171ab976705899ed, https://doi.org/10.1088/1748-0221/4/04/P04001
24. Bishop CM (2006) Pattern recognition and machine learning. Springer, Cambridge
25. Keck T (2017) FastBDT: a speed-optimized multivariate classification algorithm for the Belle II experiment. Comput Softw Big Sci 1(1):2. arXiv:1609.06119, http://arxiv.org/abs/1609.06119, https://doi.org/10.1007/s41781-017-0002-8

26. Friedman JH (2002) Stochastic gradient boosting. Comput Stat Data Anal 38(4):367–378. https://www.sciencedirect.com/science/article/pii/S0167947301000652, https://doi.org/10.1016/S0167-9473(01)00065-2
27. Efron B (1979) Bootstrap methods: another look at the jackknife. Ann Stat 7(1):1–26. http://projecteuclid.org/euclid.aos/1176344552, https://doi.org/10.1214/aos/1176344552

Chapter 4
Fast Reconstruction for the High Level Trigger

The High Level trigger (HLT) plays a crucial role for the success of an experiment, as pointed out in Sect. 2.2. Its decision logic, which defines whether a collision event is accepted or discarded, defines which physical results can be extracted from the collected data and whether a certain analysis has enough statistics to be possible at all. A sub-optimal event selection at this stage looses them irretrievably. As a first event tagging is already performed at this stage, even accepting an event but misclassifying it may lead to wrong results. Additionally, in the case of Belle II, it also determines which PXD hits are stored for later offline reconstruction. With the PXD hits, the resolution of many important and physically interesting parameters improves significantly (see also Chap. 6).

Besides the high impact on the physical results, there are other requirements that the HLT must fulfill. The most limiting is the processing time per event, which defines the throughput of the trigger stage. The number of HLT server nodes together with the expected output rate of the Level 1 (L1) trigger imposes conditions on the runtime per event. As described in the following sections, this requirement can only be fulfilled with a smart approach for the HLT reconstruction. A first implementation was developed during this thesis.

This chapter describes the novel proposal for the HLT reconstruction, which is called *FastReco*. After an introduction in Sect. 4.1, the runtime requirements are discussed together with detailed processing time measurements in Sect. 4.2. The principles of the newly developed approach and the implementation of a more general software trigger setup are presented in Sect. 4.3. The chapter concludes with an overview on technical details of the new HLT multi-core processing framework that was developed during this thesis (Sect. 4.4) and which is currently evaluated for its use in the data acquisition of Phase 3.

N. Braun, *Combinatorial Kalman Filter and High Level Trigger Reconstruction for the Belle II Experiment*, Springer Theses,
https://doi.org/10.1007/978-3-030-24997-7_4

4.1 Introduction

The overall performance of the HLT is characterized by the following quantities:

- The efficiency ε_α for each decay event topology the e^+e^- collision produces (also called *channel*) α,
- The final output rate of the software trigger stage which is passed to the following storage nodes

$$R = \sum_\alpha \varepsilon_\alpha i_\alpha \, ,$$

  with i_α being the input rate passed into the HLT by the L1 trigger for the channel α and
- The total processing time per event

$$T_\Sigma = \sum_\alpha w_\alpha t_\alpha \, ,$$

  with t_α the processing time per channel α of a single event and w_α the fraction of this channel in the set of events the HLT has to process. The fraction w_α is related to the L1 input rate by

$$w_\alpha = i_\alpha / \sum_\alpha i_\alpha \, .$$

The conditions for ε_α, T_Σ and R will be discussed in the following. But first, the input rate i_α is determined from the L1 performance.

4.1.1 Event Composition

Due to the accelerator revolution frequency, a positron and an electron bunch collide every 4 ns at the IP of the Belle II detector. Not every collision produces a physically interesting topology and also not every interesting topology generated during these collisions can be detected in the acceptance range of the detector. Only those decay topologies which are detectable at Belle II can be triggered in the L1 trigger. For calculating the input rates for the trigger, only the so called *fiducial cross section* is of interest, which factors out the insensitive detector regions. In Table 4.1, the fiducial cross sections for the channels with the largest contributions are given for the SuperKEKB default center of mass energy of 10.573 GeV. They are mainly taken from [1] or from the calculations of the applied generators directly. The generators are also described in [1]. All other channels are not discussed here as their cross section is negligible.

Using the anticipated nominal instantaneous design luminosity of SuperKEKB (see Sect. 2.1) of

Table 4.1 Studied channels with their fiducial cross sections and rates including the applied cuts for each channel. The cuts to the acceptance region of the detector are chosen to only produce detectable events. If cuts deviating from [1] are used, the cross sections are taken from the calculations of the generator directly and the given uncertainty results from MC systematic

Physics process	Cross section σ_α in nb	Rate R_α in kHz	Generator	Generator cuts (Notation taken from [1])
B^+B^-, $B^0\overline{B}^0$	1.05 ± 0.10	0.84 ± 0.08	EvtGen	see [1]
$u\overline{u}(\gamma)$	1.61	1.28	KKMC, PYTHIA8	see [1]
$d\overline{d}(\gamma)$	0.4	0.32	KKMC, PYTHIA8	see [1]
$c\overline{c}(\gamma)$	1.3	1.04	KKMC, PYTHIA8	see [1]
$s\overline{s}(\gamma)$	0.38	0.30	KKMC, PYTHIA8	see [1]
$e^+e^-(\gamma)$ (larger angle)	123500 ± 3400	98800 ± 2700	BHWIDE	$0.5° < \theta^* < 179.5°$, see text
$e^+e^-(\gamma)$ (smaller angle)	298 ± 7	238 ± 6	BHWIDE	$10° < \theta^* < 170°$, see text
$\gamma\gamma(\gamma)$	4.92 ± 0.15	3.94 ± 0.12	BABAYAGA.NLO	$10° < \theta^* < 170°$, $E_{\text{min}} > 150\,\text{MeV}$
$\mu^+\mu^-(\gamma)$	1.148	0.918	KKMC	see [1]
$\tau^+\tau^-(\gamma)$	0.919	0.735	KKMC	see [1]
$\pi\pi(\gamma)$	0.16	0.13	Phokhara	see [1]
$e^+e^-e^+e^-$	39.3 ± 0.5	31.5 ± 0.4	AAFH	$W_{\ell\ell} > 500\,\text{MeV}$
$e^+e^-\mu^+\mu^-$	15.7 ± 0.4	12.5 ± 0.3	AAFH	$W_{\ell\ell} > 500\,\text{MeV}$

$$\mathcal{L} = 8 \times 10^{35}\,\text{cm}^{-1}\,\text{s}^{-2} = 0.8\,\text{nb}^{-1}\,\text{kHz}$$

one can calculate the rates of the different channels using

$$\dot{N}_\alpha = \mathcal{L} \cdot \sigma_\alpha\,,$$

which is also included in the table. Not all of the channels in the table are of interest for the later analyses. The two-photon channels $e^+e^-e^+e^-$ and $e^+e^-\mu^+\mu^-$ will be treated as uninteresting and tagged as background in the following. Bhabha scattering has the highest cross section and is also not interesting for later analyses. However, this well understood channel can be used for luminosity determination as well as particle identification or tracking studies, so a prescaling with a known factor is necessary.

The table shows two entries for the e^+e^- channel. The acceptance range of the detector is smaller than both angle cuts. From a naïve point of view, both entries should therefore lead to the same result since the additional events in the larger angle category are not detectable anyways. Due to the boost, particle showers, radiation and multiple scattering, even the e^+e^- with a small angle θ^* in the CMS produce entries in the ECL in a significant amount of cases. It will become apparent later, that the current implementation of the L1 trigger struggles to cut away especially those events.

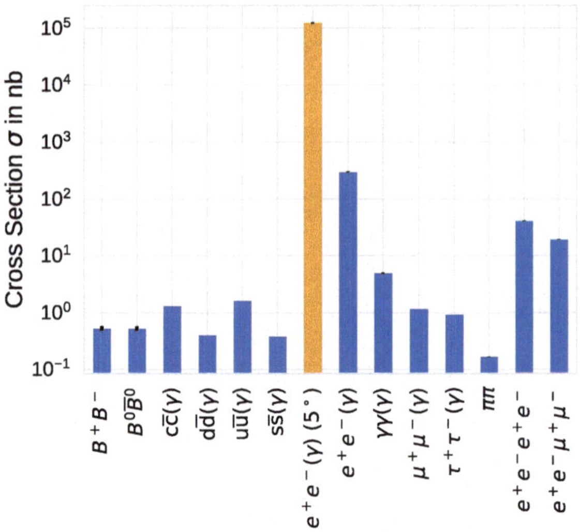

Fig. 4.1 Fiducial cross sections before L1 selection of the most important channels. Shown are the results of Table 4.1 for better visualization. The vast majority of events are uninteresting topologies such as e^+e^- or two photon processes. The additional entry of Bhabha scattering for low angles is marked orange

As can be seen in Fig. 4.1, the majority of events in the fiducial region are made up from uninteresting processes. The first step is therefore to apply the hardware L1 trigger described in the following.

4.1.2 Level 1 Trigger

The implementation of the L1 Trigger is discussed in more detail elsewhere [2] and is not within the scope of this thesis. The most important parts are a simplified CDC tracking [3] in two and three dimensions and a fast ECL reconstruction [4]. The interesting aspect of the hardware trigger for this thesis is its performance and efficiency for the different channels. To this end, the channels according to Table 4.1 are simulated using the nominal detector setup of Belle II with beam-induced background anticipated for the full instantaneous luminosity. The current implementation of the L1 trigger simulation [5] is used to calculate the efficiencies of the hardware trigger. It uses an implementation of the trigger algorithms in software but does not necessarily give the same results as the hardware implementation. In the following, the shown simulation is assumed to be the final L1 trigger rate. Figure 4.2 summarizes the efficiencies of the different channels of the L1 trigger decision.

As discussed in Sect. 2.2, the L1 trigger should deliver an average rate of 20 kHz (maximal 30 kHz) to the HLT. Summing up the different channels and including the large angle e^+e^- process, the final output rate of L1 is (382 ± 16) kHz, which is clearly higher than expected. Additionally, the implementation has inefficiencies in many interesting low-multiplicity channels, e.g. $\pi\pi$ or $\gamma\gamma$. Therefore, the L1 simulation in its current form cannot be used to supply the input for the HLT studies presented in

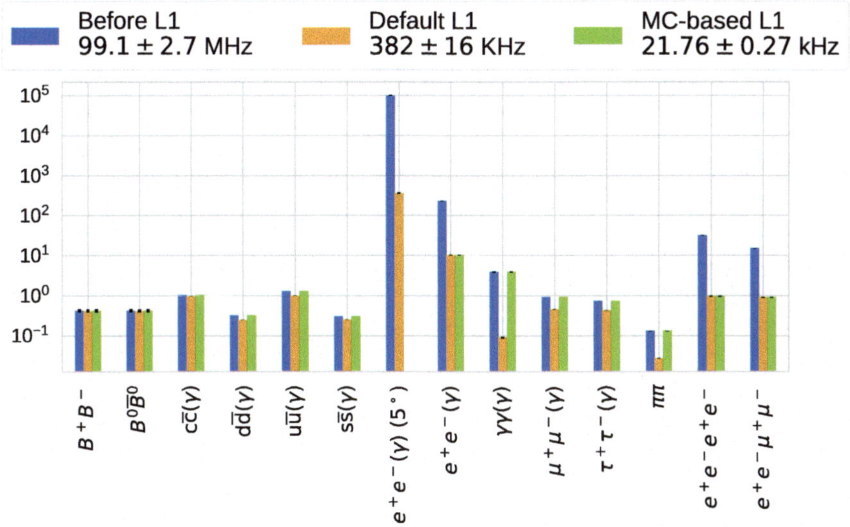

Fig. 4.2 Output rates before and after the L1 decision for the different channels. The default L1 trigger decision was extracted using a software simulation of the trigger. Since it did not reach the final required output rate of approximately 20 kHz, an artificial second implementation using MC information was used instead

the chapter. Instead, it is replaced by an artificial trigger stage using the MC truth information of the generators:

- The efficiency of the physically interesting channels B^+B^-, $B^0, \overline{B}^0$, $\gamma\gamma$, $\mu^+\mu^-$, $\tau^+\tau^-$, $\pi\pi$ and the continuum channels is set to 100%,
- The e^+e^- events with small $\theta^* < 10°$ or large $\theta^* > 170°$ are dismissed and
- The remaining efficiencies ($e^+e^-\mu^+\mu^-$, $e^+e^-e^+e^-$, remaining e^+e^-) are taken from the L1 simulation.

The resulting MC-based L1 trigger rate is also shown in the figure. The final rate of (21.76 ± 0.27) kHz fulfills the requirements described in Sect. 2.2 and events surviving the trigger stage can now be fed into the HLT reconstruction.

4.1.3 Requirements

With the simulated data sizes, the planned bandwidth and the anticipated nominal instantaneous luminosity, the overall output rate of the HLT should not be higher as $R = 10$ kHz or 12.5 nb. This requirement is mainly determined by the storage size and speed. To fulfill this, the average acceptance rate of the HLT decision must be in the order of 2/3. Since the efficiencies for signal channels, e.g. BB, continuum or $\tau^+\tau^-$, are expected to be in the order of 100%, the only possibility to regulate the output rate is to scale the background channels $e^+e^-e^+e^-$ and $e^+e^-\mu^+\mu^-$ as well as the Bhabha channel $e^+e^-(\gamma)$.

The maximum processing time per event T_Σ depends on the processing power of the HLT farm. At the time of writing, the final HLT size is still undecided and is dependent on the background conditions and the instantaneous luminosity achieved by the accelerator. For studying the feasibility of the trigger setup, in the following a setup of 20 HLT units will be assumed, which is currently the planned design in its full expansion stage. Each HLT unit will consist of 16 workers and 20 cores per worker (see also Sect. 2.2). Summing up the number of available CPUs and the input rate of 20 kHz (30 kHz), the maximum time per event is $T_\Sigma = 0.32$ s (0.213 s). If the final HLT decision is not made during this time in a large number of cases, the data cannot be read out fast enough from the trigger system and is irretrievable lost.

A first countermeasure is obviously to increase the number of HLT worker nodes. Apart from the high financial costs, this also brings the downside of additional networking, higher probability of failure and larger need for caching in e.g. the PXD readout boards. The other method to comply with the requirement is to keep the reconstruction time per event small. The next sections will discuss, if and how this can be achieved.

4.2 Processing Time Studies

The major requirement for the HLT reconstruction discussed in this thesis is the processing time, which, as shown in the last section, is constrained by the output rate of the L1 trigger and the number of available processing cores. For N events running on p cores in parallel on one HLT worker unit, the total runtime can be written as

$$T = T_{\text{init}} + N \cdot T_{\text{work}}(p) + T_{\text{terminate}} \ .$$

As the initialization and terminate time is expected to be negligible compared to the time needed for the actual work, the following studies only start measuring the time after a "burn-in" time of 100 events and stop the measurement when there are only 100 events left. The work time T_{work} is briefly referred to as T.

The time spent on the actual work of the process can be split up in the time for the calculation on a single processor $T(1)$ and the *speedup* function $\eta(p)$ depending on the number of processes

$$T(p) = T(1)/\eta(p) \ .$$

Both terms are discussed separately below. Some preliminary results can also be found in [6].

4.2.1 Speedup $\eta(p)$

Figure 4.3 shows the measured speedup as a function of the number of utilized worker processes p for typical B^+B^- events with beam-induced background. The speedup is hereby determined as the ratio between the reconstruction time with a single process[1] and the time with multiple processes. The number of processes p depicts the number of worker processes; not including the additional input and output process. As described, the times are measured from event 100 until $N - 100$ to not be influenced by the initialization or termination phase. All the studies are performed on parts of the HLT hardware accessed between the data acquisition of Phase 2 and Phase 3. Important technical details of the utilized machine are summarized in Table 4.2. Some parts of the information are explained in this section.

From a naïve point of view, each added worker process should give the same amount of additional working power, so a speedup of $\eta(p) = p$ could be expected at first. It can clearly be seen, that the measured speedup does not correspond to this expectation. A more correct description of the speedup of multiprocessing applications is given by Amdahl's law [8]

$$\eta(p) = \frac{1}{(1 - r_p) + \frac{r_p}{p}},$$

where r_p denotes the fraction of the application which is calculated in parallel. Under ideal conditions, this fraction is 100% and the naïve speedup of $\eta(p) = p$ is restored.

However, the data can be better described with $r_p = 99\%$ for $p < 20$. The task of the remaining section is to explain why not 100% of the reconstruction work can run in parallel. Multiple reasons can lead to a deviation between the theoretical speedup and the practical measurements and it is in general hard to find out all effects. A limited subset of possible problems is discussed in the following. Using toy studies in a simplified environment, the different effects are studied for their influence on the scaling behavior of the HLT reconstruction. Although not all discussed effects influence the shown measurement, they can be used to limit the scope of the measurements. In the end, only $\eta(p = 40)$ is of interest for the final HLT setup, since all HLT worker CPUs will be utilized.

4.2.1.1 Effect of Input and Output Processes

In Sect. 2.3, the multiprocessing architecture of basf2 was discussed. Two additional processes—the input and output process—are needed in addition to the p worker

[1]There is a subtle difference between the reconstruction with a single process and single-core processing as introduced in Fig. 2.8. In this thesis, only the worker processes are counted, so the reconstruction on a single process uses actually three cores: one for the worker, one for the output and one for the input process (cf. Sect. 2.3). All the studies in the following are normalized to this case.

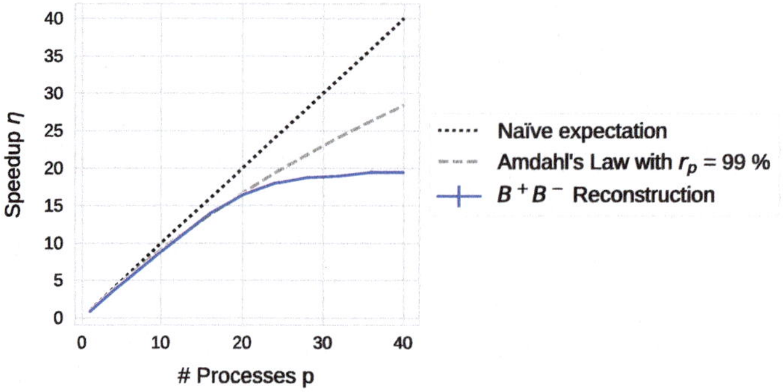

Fig. 4.3 Measured speedup for simulated BB events with full luminosity beam-induced background. The measurements are performed on a single HLT worker node. It can be clearly seen, that the measured speedup does not comply with the expectation of $\eta(p) = p$

Table 4.2 Relevant technical details of the utilized HLT worker nodes for this study. Some of the figures are explained in more detail along with the presented study. Data taken from [7] or directly from the machine specifications

HLT worker node	
CPU Sockets	2
Physical CPU cores per Socket	10
Threads per core	2
CPU Model	Intel Xeon
CPU Version	E5-2660 v3
Base Frequency	2.60 GHz
L1 cache (each CPU)	32 kB
L2 cache	256 kB
L3 cache (each Socket)	25 600 kB
RAM	47 GB

processes performing the reconstruction. Both the input as well as the output is a sequential process. The runtime of both processes can influence the scaling behavior of the multiprocessing application. Figure 4.4 shows a schematic drawing of the performed work on different processes for an exemplary task to reconstruct nine events with three worker cores. The shown cases are the two borderline cases of Amdahl's law for $r_p \to 0$ and $r_p \to 1$. Using the notation of the process names "input", "reco" and "output", this corresponds to

$$T = \begin{cases} \frac{1}{p} \cdot T_{\text{reco}}, & \text{if } T_{\text{reco}} > (p-1) \cdot T_{\text{input}} \\ T_{\text{input}}, & \text{otherwise} \end{cases} . \qquad (4.1)$$

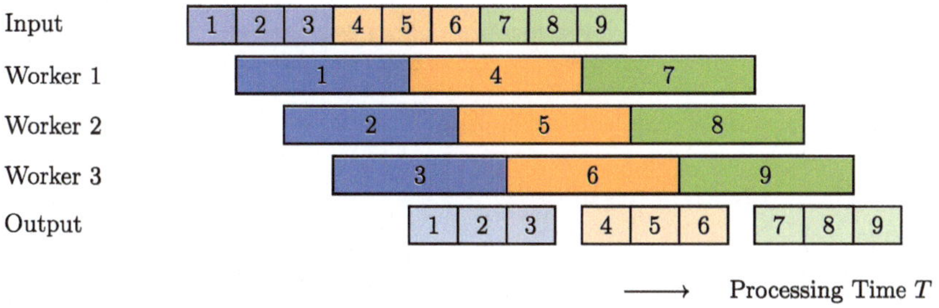

(a) Case, in which the reconstruction time T_{reco} is large compared to the input time ($T_{\text{reco}} > (p-1) \cdot T_{\text{input}}$). For a large number of events the total reconstruction time is dominated by the time needed for the reconstruction, as there is no waiting time.

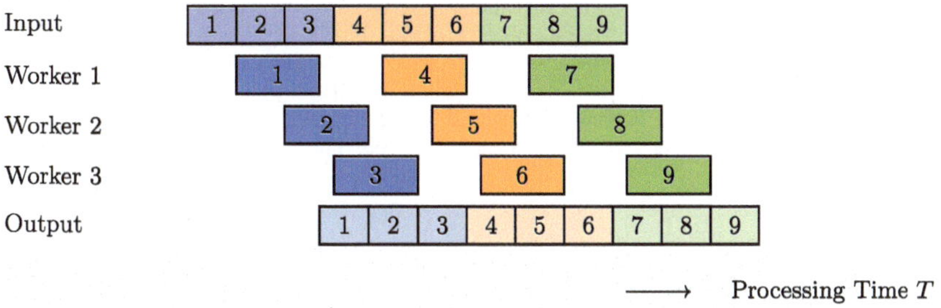

(b) Case, in which the reconstruction time T_{reco} is comparable to the input time ($T_{\text{reco}} \lessapprox (p-1) \cdot T_{\text{input}}$). As the worker processes have to wait for the input process to deliver new data (white areas), the total reconstruction time is dominated by the sequential input process and does not scale with the number of processes anymore.

Fig. 4.4 Exemplary processing times of nine events distributed over three worker processes. Two different borderline cases of Amdahl's law are shown. As the events must be sent from the input to the worker and from the worker to the output processes, there is the need for synchronization and wait times. The influence of streaming and serialization is not shown

A similar formula also applies to the output time T_{output}, but is omitted here for brevity. This means a slow input process can lead to sequential parts in the calculation which worsens the scaling behavior.

To visualize the formula and test the general multiprocessing framework, simplified measurements are performed and shown in Fig. 4.5. An artificial reconstruction path with modules, that just wait a certain amount of time, is processed with different numbers of workers. The wait time of the modules in the input, reconstruction or output path can be adjusted. To not bias the input module due to disk operations, trivial event data consisting of blocks of random data are generated on the fly, which is fast enough compared to the chosen input wait times.

The times for the input and output modules T_{input} and T_{output} are set both to 10 ms. The reconstruction time T_{reco} is chosen to be 20 ms, so that the transition between the two cases of Eq. (4.1) happens at $p = 2$. To reduce the influence of external effects on the computer, each measurement is repeated 3 times.

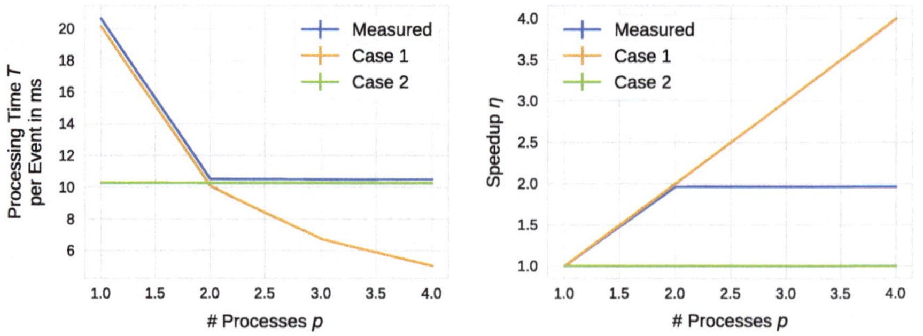

Fig. 4.5 Measured processing time for an artificial example of a path, which just wait a certain amount of time during input, reconstruction and output. The sent event data is small and trivial to (de)serialize. The expected processing times according to the two border cases discussed in Eq. (4.1) are also shown. The wait times are chosen so that the transition between the two cases happens at $p = 2$, which is clearly visible in the figure. The right plot shows the expected and measured speedup η for the different cases. In case 2, the speedup is constant, which is non-optimal as additional cores do not increase the performance. The uncertainty is too small to be visible

The data points follow the two descriptions for the full processing time perfectly in the regions where are applicable. This shows, that the basf2 multiprocessing framework is in principle able to scale with the number of processes in this simple toy example if the input process is fast enough. Previous measurements have already shown that the input stage of the HLT server farm is capable of streaming the events fast enough to the machines [9].

In contrast, the measurements shown in the following do not use this input stage, but read the presimulated events from disk. The input data and a copy of the detector conditions database are stored on a RAM disk for faster access. The disk is a special form of a temporary file storage facility (tmpfs), created by the unix operation system, which makes the input data directly accessible in the RAM. No output is stored to disk at the end of the calculation. In total, the time spent for disk operations in the input process is below 1 ms, which can be determined in single-core measurements. As the typical reconstruction time is larger than 100 ms, the disk operation in the input process does not cause any sequential part of the reconstruction and therefore cannot explain the measured speedup.

4.2.1.2 Serialization

A slow reading of the events into the HLT node (either from disk or from the event builder) is not the only effect that worsens the input time and the overall scaling. Additional deteriorating effects on the scaling behavior and the total processing time can come from the (de)serialization of the event data before and after sending it to the ring buffer.

Serialization takes place at the end of the input and each worker process, deserialization at the start of the worker and output processes. A large contribution of the serialization can therefore alter the ration between T_{input} and T_{reco}, as the serialisation must be performed p times more often in the input process. Figure 4.6 shows the impact of the serialization on the full processing time and the scaling with different input files and reconstruction lengths. A complex data store content with approxi-

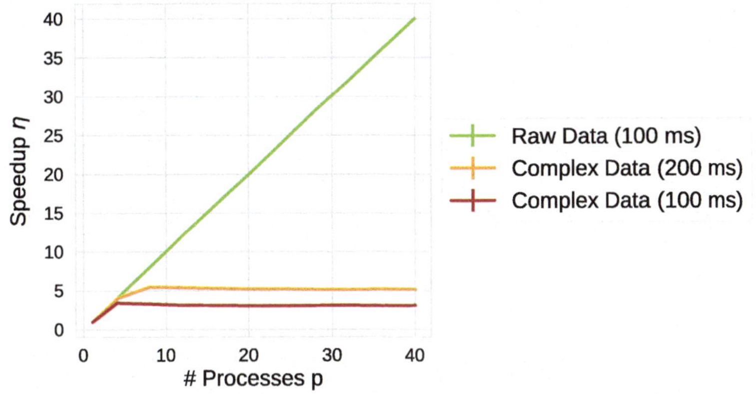

(a) Speedup for different reconstruction times.

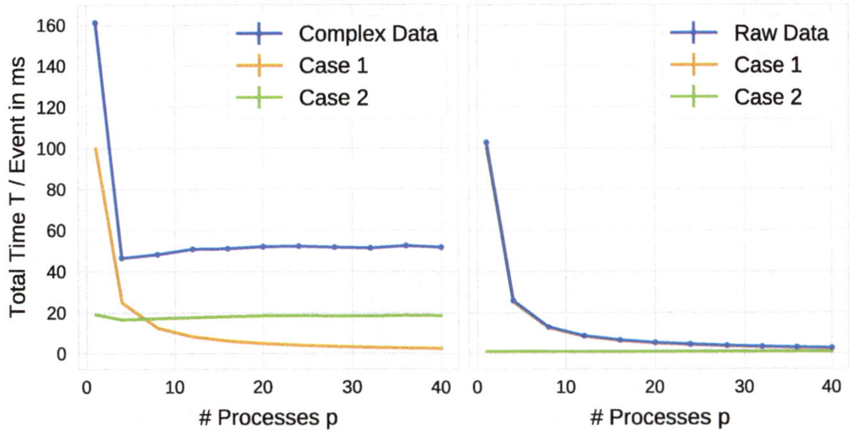

(b) Total processing time per event for a reconstruction time of 100 ms including the predictions using Equation (4.1).

Fig. 4.6 Measured speedup (top) and total processing time (bottom) for different cases of input data. The reconstruction time is simulated with a sleep. The complex case (bottom left) looks similar to the artificial example in Fig. 4.5. The difference between the scaling behavior of the complex and the raw data can not only be explained by the longer time it takes to read the events from disk, as it would then follow the green curve. It is also caused by the time needed for serialization. Using raw events (bottom right) matches exactly with case 1, which is shown in yellow mately hundred different objects and a reduced event containing only the small raw data objects is streamed in the study.

When using the simple raw data as input, the reconstruction is able to scale optimally even for the small reconstruction time of 100 ms. The total time and the speedup follow exactly the predictions of Eq. (4.1) for short input times. As no degradation is seen even for $T_{\text{reco}} = 50\,\text{ms}$ (not shown), an upper limit of $T_{\text{input}} < 50\,\text{ms}/(40 - 1) = 1.28\,\text{ms}$ for the combined time of data reading from disk and the serialization T_{ser} can be calculated. This makes it fast enough for any reasonable reconstruction.

The picture changes when using more complex event data. Due to the larger file size, the disk operation needs a longer time of approximately $T_{\text{disk}} = 18.67\,\text{ms}$. As shown in the lower part of Fig. 4.6, this alone cannot explain why the speedup breaks down at approximately $p = 4$. Assuming that for $p = 40$ the total processing time is, according to Eq. (4.1), only determined by the time spent in the input process, the time required for serialization is

$$T_{\text{ser}} = T(40) - T_{\text{disk}} = (34.07 \pm 0.34)\,\text{ms}$$

and therefore significantly larger for the complex data.

The toy example shows, that serializing complex event data can have a significant impact on the speedup and the total reconstruction time. As the HLT only has to serialize simple raw objects, its speedup as well as the total reconstruction time is not influenced by this. However, it plays a role when utilizing basf2 on large scale computing farms for MC event generation or reprocessing of the recorded data [10].

4.2.1.3 Streaming

The effect of the data streaming between the processes is more complex than just an additional contribution to the processing times, as in the case of the serialization. The ring buffer used in the current multiprocessing implementation has a fixed size. This means that if event data does not fit into the ring buffer anymore because the other processes did not process the data fast enough, the writing process needs to wait until a slot is free again. Additionally, every writing and reading step needs to lock the ring buffer for exclusive usage. Section 4.4 shows possibilities how to circumvent this problem and also other drawbacks of the ring buffer implementation.

In Fig. 4.7, the processing time and the scaling behavior for different streamed data store sizes are shown. To avoid mixing the influences of serialization and streaming, the data used consists only of directly produced, simple random data instead of complex event data read from disk. The produced event data sizes are much smaller than the RAM size of the machine used. Therefore it is not expected that the memory allocation has an impact on the measured processing times.

For a large range of different data sizes the measured processing times T stays the same. At very large input sizes (e.g. 30 MB) in the order of the ring buffer capacity of 60 MB, the ring buffer is filled up faster. As the input and worker processes have to

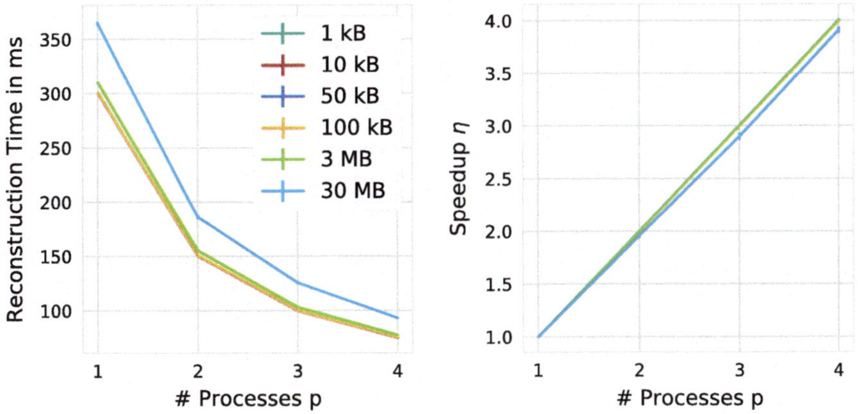

Fig. 4.7 Measured processing time of the example path with only sleeper modules using artificial generated event data of different sizes. Up to the expected event sizes of 100 kB, there is no visible impact on the measured time. For large event sizes, which can occur during offline reconstruction where the multiprocessing framework is used as well, the data store streaming starts to play a deteriorating role

wait for a free slot before filling in the data, this synchronization leads to a measurable delay especially in the input step, which grows with a larger number of processes. This leads to a worse speedup compared to smaller data samples. With nominal Phase 3 instantaneous luminosity, the total data size is expected to stay in the order of 100 kB, where no large influence of the streaming is seen. Again, this effect cannot explain the measured speedup in Fig. 4.3.

4.2.1.4 Measurement Module

The shown processing time measurements were performed by adding an additional module to the path, which stores the wall time clock after 100 and $N - 100$ events and calculates the difference. The module is added to the path, which is used in the output process. Although the task of this module is narrow, starting the module itself in every event and checking the event number can in principle have an influence on the processing time. In Fig. 4.8 the total clock time of a full execution of different scenarios is shown with and without the measurement module. For later reference, the detailed module-time measurement required in the next section is also compared. As expected, the influence of the different measurement modules is smaller than all the anticipated reconstruction times and therefore negligible.

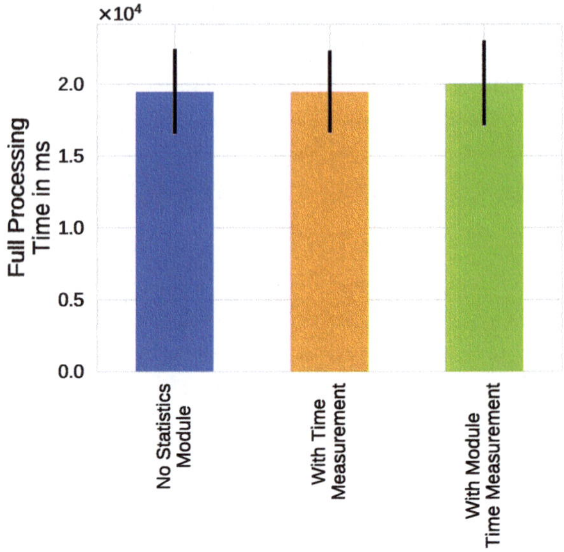

Fig. 4.8 Time required for the full processing of an artificial path with small reconstruction and input/output waiting times of 20 ms and 10 ms respectively. Even in the case of this small total processing time the measurement modules play no role. The module-time measurement stores the reconstruction time of each reconstruction module in addition to the full processing time and is used in the next section to determine the times of each reconstruction step

4.2.1.5 Hyper-Threading

Apart from the influences of the multi-core processing implementation described in the section before, also the computing infrastructure and environment can have an influence on the processing time.

As discussed in Table 4.2, the utilized HLT worker nodes have 20 physical cores. A physical core consists of a single set of components, e.g. one arithmetic unit, one cache etc. Using a technology called hyper-threading [11], which is a form of simultaneous multithreading developed by Intel [12], each physical core is matched to two virtual cores. Every component is shared between the two virtual cores except the section storing the control and general purpose registers. This allows the operating system to schedule two independent processes that are alternatingly processed by the shared resources. The stall times of one process due to data dependencies, branch misprediction or other external waiting times can be used by the other process running on the same physical core. As these "additional" cores do not have the same processing power as a real core, the expected additional speedup is only up to 30% per additional process [11]. Figure 4.9 shows the measured speedups for the reconstruction of B^-B^+ events (repeated from Fig. 4.3) with the updated expected scaling. Using Amdahl's law separately in the two parts $p < 20$ and $p > 20$, the data can be described well with

$$r_p\big|_{p<20} = 99\% \qquad r_p\big|_{p>20} = 78\%\ .$$

Although hyper-threading explains the overall shape, it still cannot explain the deviation between the expected 100% and the measured values.

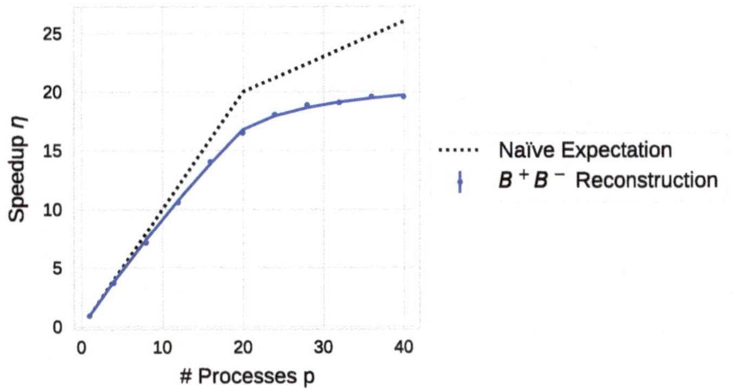

Fig. 4.9 Updated version of Fig. 4.3 with the new expectation including hyper-threading. The measured data is shown as points (error bars are too small to be visible). The measured speedup clearly shows an influence by this effect, although the data is still best described with non-optional fractions of parallel processing $r_p \neq 1$ (shown as solid line)

4.2.1.6 CPU Cache

The input data as well as temporary calculation results of the reconstruction are stored in the main memory of the worker nodes. Although the access is much faster than any disk operation, it is slow compared to the length of a CPU cycle (see Fig. 4.10). To circumvent long stall periods of the CPU, a system of caches is used to buffer each read and write operation to the main memory. A typical hierarchy of caches is shown in Fig. 4.10 together with common latency times for the read access. If the data cannot be found in a cache, it has to be accessed in the next stage of the memory hierarchy with additional waiting times. It is therefore beneficial to have as few cache misses as possible.

The number of cache misses can be measured with the `perf` tool [14], which has little to no influence on the reconstruction itself, as it uses the performance counters of the CPUs which are filled anyways. In Fig. 4.11 the number of counted L1 cache misses for the reconstruction of 5000 B^-B^+ or e^-e^+ events is shown. As the e^-e^+ events are smaller and less complex, a larger fraction of the input and temporary data fits into the L1 cache inducing less cache misses in total. The cache miss rate per instruction is mainly influenced by the application code and the memory scheme of the program. It is therefore comparable in both event types with approximately 16 L1 cache misses per 1000 instructions for $p < 20$. Beyond that, the number of misses and the miss rate increased rapidly if more than one process is scheduled per physical core. The reason is that the two virtual CPUs share a single L1 cache instance on each physical core and induce a large amount of additional cache misses. For each cache miss, an additional latency is induced, which causes the non-optimal scaling above $p > 20$.

For $p < 20$ the number of cache misses is relatively constant and in first order no additional speedup penalty should be caused. If a date is not found in the L1, it has to

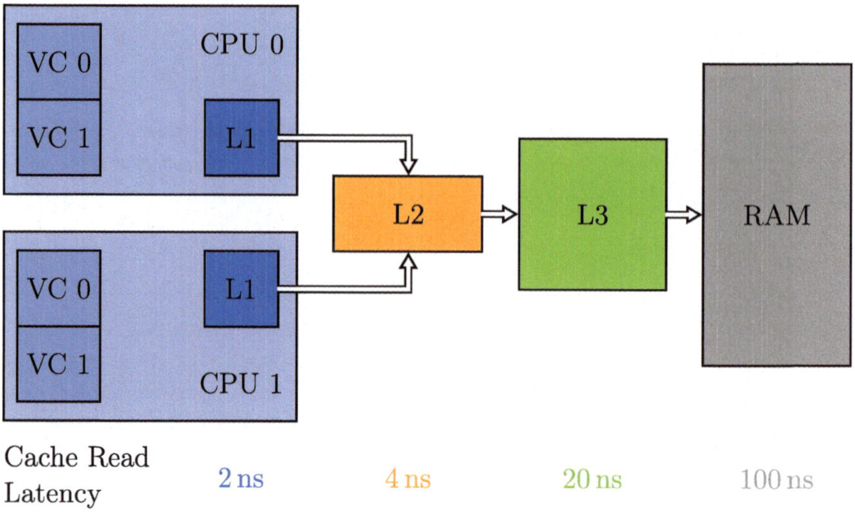

Fig. 4.10 Typical cache hierarchy between the CPUs and the main memory (RAM) together with approximate latencies for a read access. The cache latencies are taken from [13], but are generally dependent on the execution environment (clock frequency, memory speeds, settings etc.). They are therefore only a rough approximation

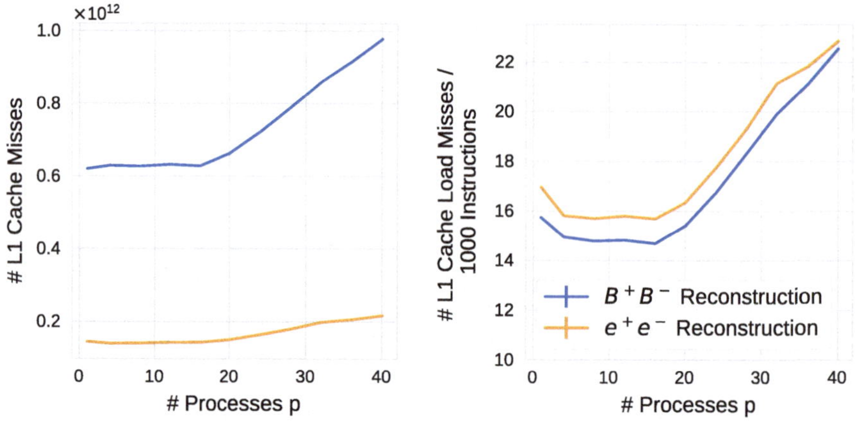

Fig. 4.11 Total number of cache load misses of the L1 data cache for different event topologies in absolute numbers and relative to the number of instructions summed for all processes. The measurements were performed with the `perf` utility. As e^-e^+ events are smaller and less complex, they do not invoke as many absolute cache misses as B^-B^+ events. The relative fraction is comparable between the event types. Beyond $p = 20$, the miss rate increased drastically

be accessed from the L2 or L3 cache. Those instances are shared between all CPUs on a single socket and each access to it is sequential. If two L1 cache misses happen at the same time, they cause an additional latency for one of the accesses. Let I be the number of instructions and m the number of L1 cache misses per instruction. The probability to have two cache misses at the same time with p cores is given by

$$P(\text{two cache misses}) = 1 - (1 - m)^{p-1} \ .$$

One of the cache misses has to wait an additional L2 (or L3) cache latency time L until it can access the cache itself. The total delay for each process is therefore given by

$$T_{\text{delay}} = \frac{1}{2} \cdot P(\text{two cache misses}) \cdot L \cdot m \cdot I \ ,$$

because on average every second cache miss ($1/2m \cdot I$) causes a latency of L with the probability P. As m is small, this can be approximated to

$$T_{\text{delay}} = \frac{LmI}{2} \left(1 - (1 - m)^{p-1}\right) \approx \frac{Lm^2 I}{2}(p - 1) \ .$$

Inserting this into the formula of the speedup, one gets

$$\eta(p) = \frac{T(1)}{T(p)} = \frac{T(1)}{\frac{T(1)+T_{\text{delay}}}{p}} \approx \frac{1}{\frac{1-\alpha}{p} + \alpha} \ ,$$

where α is given by

$$\alpha = \frac{1}{2} \frac{I m^2 L}{T(1)} \ .$$

Compared to Amdahl's law, it is clear that the additional synchronization delays caused by the simultaneous L1 cache misses induce a sequential fraction of the calculation. Averaging the latency time to $L = 10$ ns for L2 and L3, the contribution of the L1 cache misses to the sequential fraction of the calculation is given by

$$\alpha = 0.53\% \ .$$

This first back-of-the-envelope calculation is in good agreement with the $r_p = 99\%$ found in the measurements. The non-optimal scaling can therefore be explained by the simultaneous cache misses inducing additional waiting times on multiple cores.

The problem is even more serious because, e.g., the shared L3 cache has a finite size. If multiple CPUs start to request data from the cache, the cache runs full faster leading to additional cache misses compared to the single-core execution. Figure 4.12 shows the miss rate per number of requested cache loads for the L1 and L3 cache. As the L1 caches are unique for each physical core, the miss rate does not depend on the number of processes for $p < 20$. The L3 cache miss rate increases linearly until it reaches its maximum around $p = 20$. Beyond that, the additional L1 cache misses cause more L3 loads of the same data and the L3 miss rate stays approximately constant. Each additional L3 cache miss compared to the single-core execution causes

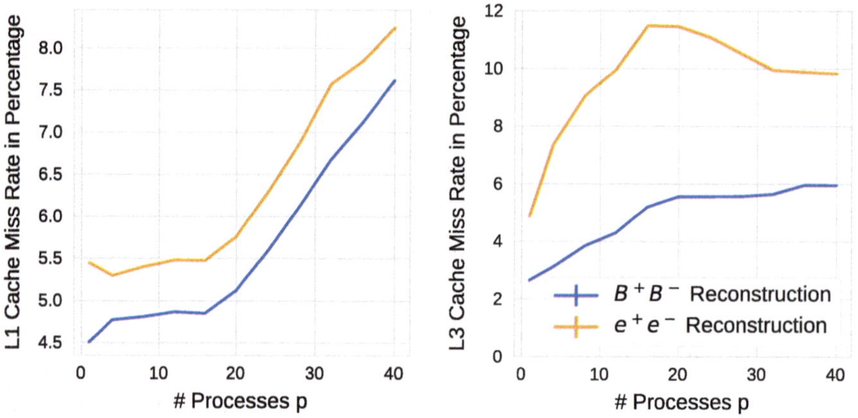

Fig. 4.12 Measured L1 and L3 cache miss rates per cache loads for a reconstruction task of 5000 B^+B^- or e^+e^- events summed for all processes. The results were recorded using the `perf` program on an HLT worker node

an additional delay which leads to a significant decrease in the scaling quality as observed in the data.

4.2.1.7 First Summary

Taking into account the computational environment and the CPU specifications, the measured scaling behavior can be explained. The cache coherence of the reconstruction software can be improved in principle, but rewriting large parts of the application is a major task way beyond the scope of this thesis. In summary, Fig. 4.13 shows the speedup achieved for various channels on the HLT machines when utilizing 40 worker processes. For all studied decay topologies, a speedup of 20 is reached. Therefore, this number is later used to calculate the processing power of the HLT machine.

Certain limits were deduced while measuring this speedup. The data used in serialization and streaming should not get more complicated or larger as this would worsen the scaling behavior. Additionally, the processing time crucially depends on an input process delivering the input data fast enough. Any additional delays caused by network bandwidth etc. can deteriorate the scaling. Also, the number of cache misses should not increase or otherwise more parts of the application cannot run in parallel. Last but not least, every additional source of sequential execution, such as logging, database access or monitoring must be performed as fast as possible.

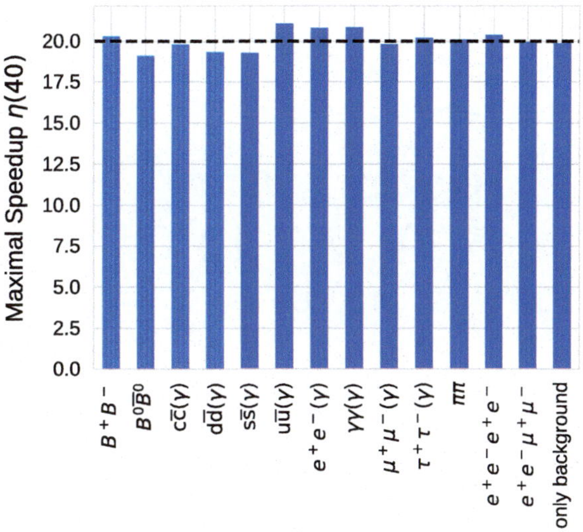

Fig. 4.13 Maximum speedup of different channels measured on the HLT worker nodes. For all channels a speedup of 20 is reached (within statistical fluctuations)

4.2.2 Single-Core Reconstruction Time $T(1)$

The second part of the final full processing power of the HLT worker units is the average reconstruction time for each reconstructed input channel on a single process $T(1)$. For the measurement, simulated data in different channels including beam-induced background is used. In each event, the execution time for every module is measured and averaged over 1000 events. The initialization and termination phase are hereby excluded. The reconstruction follows the default module path used for offline reconstruction without the time-consuming multiple hypothesis fitting, the ECL cluster recovery for clusters from bremsstrahlung and the vertex fitting. Currently, neither of these information is used for the final HLT trigger decision. Especially the vertexing may be needed at some point in the future to tag the calibration channels.

The modules are grouped according to their different tasks. Figure 4.14 shows an exemplary result. The work performed in this thesis—the event time (T_0) determination as well as the tracking improvements with the CKF—are already included in this measurement.

The largest contribution to the reconstruction time comes from the track fitting. Fitting—mainly the track fit—also has a large contribution to many of the remaining task groups: during SVD tracking a fit of the already found CDC tracks needs to be performed (see Chap. 6), the event time extraction needs to refit the tracks multiple times (see Chap. 5) and, of course, the final track fit itself is also influenced by the fitting speed. The reason for the track fit needing a relatively large fraction of the total processing time is the complex material and geometry handling required during the track extrapolation (see Sect. 3.3). This extrapolation also plays a crucial role in the ECL track matching and increases the time needed for this step. Many high energy physics experiments suffer from a slow extrapolation (see e.g. [15]). In many cases,

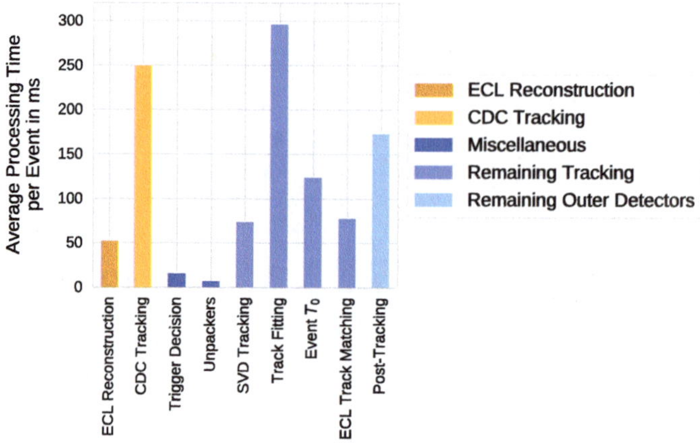

Fig. 4.14 Average measured processing time of the online reconstruction software executed on 1000 B^+B^- events with beam-induced background

this can be solved by a simplified geometry, which is better suited for access during extrapolation and fitting. Also for Belle II there has been work on this topic [16], but still the fitting takes a large fraction of the time. Possible solutions for this problem may be found in the future (e.g. by utilizing ACTS [17]), but go beyond the scope of this thesis. The remaining contributions such as ECL reconstruction, the trigger decision or the unpacking of the raw data play only a minor role compared to the rest of the reconstruction.

All results for the channels discussed in the beginning of this chapter are given in Fig. 4.15. As expected, the more complex a channel is and the more decay particles it contains, the more time the reconstruction will take. Between empty events consisting only of beam-induced background and the largest BB events lies approximately an order of magnitude.

The processing time is not only highly dependent on the channel, but also on the number of particles, signal and background measurements, which can vary between events of the same decay topology. In Fig. 4.16 the average processing time together with the 50 and 90% band is shown. Fortunately, for the calculation on the HLT, only the average processing time is of interest. However, the large span of the processing times per channel shows that e.g. unexpectedly high beam-induced background may lead to large deviations of the reconstruction times compared to what is presented in this thesis.

As discussed in the previous section, the different channels are not produced with the same frequency. To calculate the processing time for an average event on the HLT, the channels are weighted by their rate after the (MC-based) L1 trigger. The results of Fig. 4.15 after the weighting are shown in Fig. 4.17. By summing up the different contributions, the average time spent on reconstructing a single event on the HLT nodes is 344 ms, which is shown in the figure as dashed line.

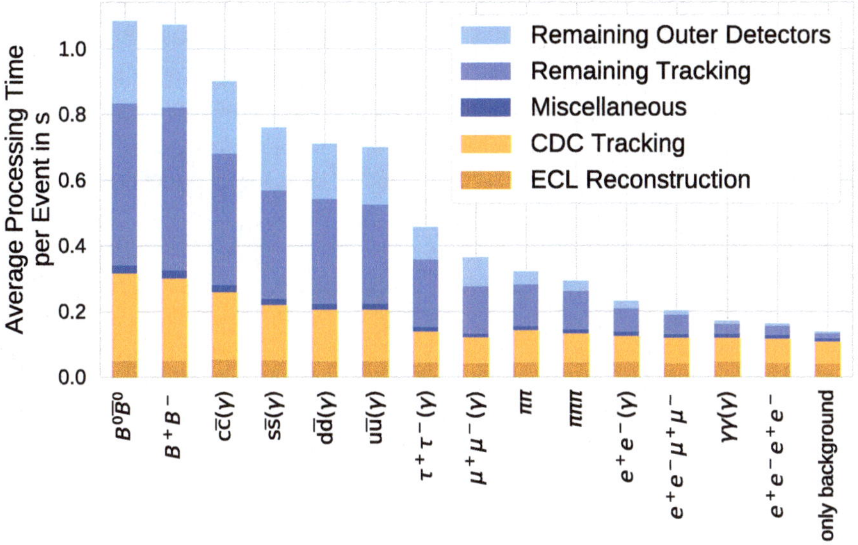

Fig. 4.15 Average processing time measured on a single core for different channels on the HLT machine split up by the reconstruction part. More complex events as B^+B^- need more time for the reconstruction. The uncertainties are suppressed for better visibility

Fig. 4.16 Median measured processing time on 1000 events with its 50 and 90% band. For every studied channel, the processing time can vary by a large factor (around 100%), which is caused by variations of the input data due to the different number of measurements or final state particles

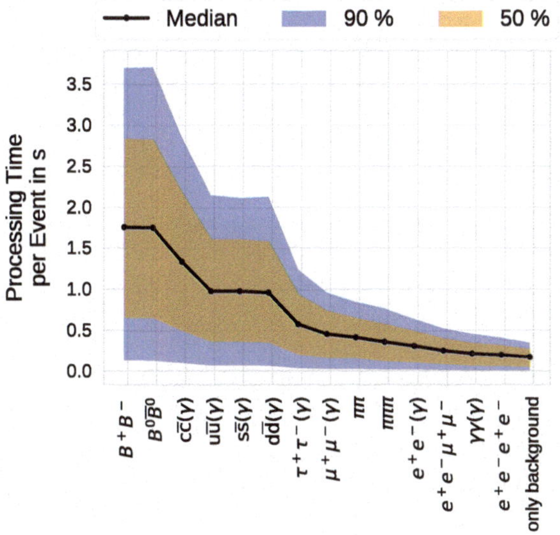

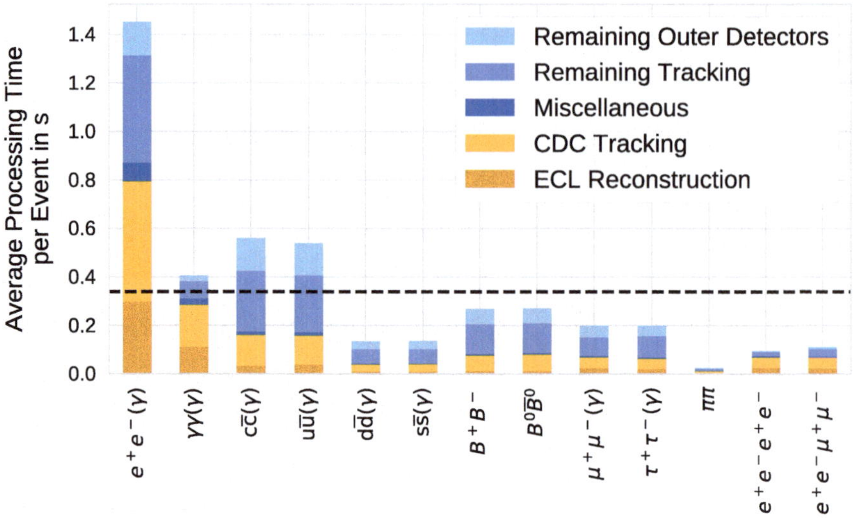

Fig. 4.17 Results of Fig. 4.15 weighted with the cross sections for each channel, so that channels with a higher cross section are upweighted with respect to the average. The average value of all displayed channels is the reconstruction time for an event with average channel mixture after the L1 selection and is shown as a dashed line

4.2.3 Summary

Using the results discussed in this chapter and the planned setup of the HLT computer farm, the expected processing power of the online reconstruction can be estimated. Hereby, some assumptions were made which are summarized in the following:

- The input stage after the event builder is fast enough to deliver the events, which has already been tested with a simplified setup [9].
- The HLT worker nodes only have to calculate the result of the reconstruction. No additional sources of delays or locks, e.g. caused by logging or monitoring are introduced.
- The event file size and the data complexity are the same for simulation and data.
- The final L1 efficiencies and rates are close to the assumed quantities at the beginning of this chapter.
- The online reconstruction follows the same module path as the offline reconstruction, except that it does not include any multiple hypothesis fitting, vertexing or bremstrahlung recovery in the ECL.

With these assumptions, the maximum scaling for a full HLT worker was determined as $\eta(40) = 20$ and the single-core reconstruction time for an average event is $T(1) = 344$ ms. The final HLT farm size has not yet been determined at the time of writing and depends on the beam-induced background conditions and the instantaneous luminosity. The server farm is designed to be upgraded continuously if additional reconstruction capacity is required. Many external influences such as other reconstruction steps, more time-consuming reconstruction parts or different input data can effect these results.

In the following, the feasibility of the HLT scheme for the full design luminosity of SuperKEKB is shown, which is dependent on the number of HLT workers. To give an example, 20 HLT units are assumed in its final stage. Each unit will host 16 HLT workers with the specifications in Table 4.2. This allows the HLT farm to process the incoming events with a maximum rate of 928.7 Hz per unit or 18.57 kHz assuming 20 HLT units. This rate is below the requirement of 20 kHz output rate of the L1 trigger. With this processing power, the HLT would not be able to reconstruct the events fast enough, leading to a severe data loss. There are at least four possibilities to circumvent this problem:

Reduce the L1 Output Rate Although in principle feasible by hardening the trigger cuts, this is not possible without losing a large amount of important events for physics analyses.

Use additional HLT Input Nodes To fulfill the requirements, the HLT server farm would have to be upgraded to 22 HLT units with 16 workers each. This is not only an additional financial burden, but also complicates the HLT network setup, the need for monitoring and the failure safety. It is not clear, if the bandwidth needed for feeding the events in time into the HLT worker nodes can still be achieved in this larger setup. Especially when assuming other possible time penalties such as unexpected background sources or additionally needed reconstruction steps, this possibility may not be feasible anymore.

Increase the Maximal Scaling per Worker Node The reduced scaling of 20 compared to the maximal possible value of 26 (20 physical cores plus hyper-threading) is caused by a large amount of cache misses during the reconstruction. With careful restructuring of the memory footprint of the code and advanced software development techniques, it might be possible to achieve the hypothetical speedup. However, this task requires a major rework of the reconstruction code, which is not possible at this stage.

Decrease the Reconstruction Time Speeding up the reconstruction by optimizing parts of the software is a task similar in size as to reduce the number of cache misses. Therefore, a different approach was chosen in this thesis. The HLT trigger decision is calculated as early in the reconstruction as possible to reduce the amount of events, where the full reconstruction needs to run. This approach and the framework developed for this is discussed in the next section.

4.3 Principles of *FastReco*

As shown in the previous section, the processing of the online software on the HLT is too slow to cope with the expected input rate at full luminosity—especially with the possible effects of additional background sources or other reconstruction steps. The solution presented in this thesis is to include additional trigger stages in the control flow of the software trigger, which rejects events early without the need for a long-lasting full reconstruction. Multiple decision stages can be implemented at different steps of the reconstruction using the reconstruction information calculated up to this point. As an example, an implementation of this *continuous HLT decision* based on the current version of the HLT menu, which was developed by Christopher Hearty [18], is shown. The example includes an early Bhabha veto to dismiss this event topology before the time-consuming full reconstruction has to be run. The approach of performing multiple trigger decisions based on increasingly detailed and precise information instead of a single trigger cut after the full reconstruction has already been successfully implemented at other experiments (e.g. [19]).

4.3.1 Continuous HLT Decision

The final output of the HLT trigger includes the ROIs, the event T_0 and the trigger tags for accepted events. On the one hand, the trigger tags serve to reject uninteresting events to stay in the maximum output rate of 10 kHz. On the other hand, they also are used to quickly sort the recorded data for the later offline reconstruction. For this, those tags also include specific calibration tags, which label events that should be used in the regular machine parameter updates (cf. Sect. 2.2). Especially these calibration tags can only be calculated after applying a full reconstruction on the recorded data. However, those tags as well as the ROIs and the event T_0 do not have to be calculated for rejected events. It is therefore possible, to abort the reconstruction early or not to start it at all, if it is already clear that the event will not be stored.

In principle, there are two possible implementations of such a pre-filtering.

- The first possibility is to use a dedicated, fast reconstruction optimized for suppression of background topologies. This was the approach taken by the former Belle experiment [20]. As a reminiscence of this former implementation, this dedicated algorithm is called *Level 3*, as this was its name at Belle.
- The second possibility implemented in this work, is to reuse fast parts of the full reconstruction. The full reconstruction is split up in different sections. After each section, cuts are introduced which decide, if the event needs to be processed by the remaining longer-running parts of the reconstruction at all. The fast parts that need to be run for the decision are called *FastReco*.

Apart from the obvious advantages of *FastReco*, e.g. much simplified code as only one reconstruction algorithm needs to be maintained, there is is also a strong motivation from the physical point of view. The Level 3 algorithm would have to be studied as thoroughly as the main algorithm in respect to systematics, efficiencies, fake rates etc. As the full reconstruction is currently used for MC production on large scales and already first studies for physics analyses are performed on the simulated data, the chance for an issue in these algorithms being solved already is quite high.

Additionally, the full reconstruction algorithm will always be treated with more care, as it defines the full offline reconstruction efficiency and resolution in the end. All systematic studies for important properties, e.g. the tracking efficiency, would need to be performed twice and the final convoluted result will be hard to understand.

Finally, it is not clear, whether the fast-running Level 3 algorithm decreases the full runtime of the software trigger at all. As a separate algorithm, this second implementation would not be able to exchange its results with the main reconstruction (or only on a very basic level), so the full reconstruction would basically have to repeat the work of the pre-stage. Assuming a rejection rate as high as $1/6$ after the pre-stage, the Level 3 algorithm would have to perform its work at least 6 times faster than FastReco, for the total processing time to be smaller. This means complex algorithms, that for example include a track fit, cannot be used. A preliminary implementation of the Level 3 algorithm ported from the Belle experiment could not achieve this rejection rate [21, 22].

As seen in Fig. 4.17, the largest part of the processing time is spent on physically uninteresting e^+e^- events, which are by far the largest fraction of events even after the L1 selection. A reasonable approach is therefore to reduce the amount of e^+e^- events early without the need to run the full reconstruction on those events.

Figure 4.18 shows the reconstruction split up in different parts and their interdependencies. The ECL reconstruction includes the creation of ECL clusters out of the bare ECL measurements and a cluster shape analysis. The details of the cluster reconstruction are described in [23]. The CDC track finding can run independently from the ECL reconstruction and has already been discussed in Sect. 3.2. As the SVD

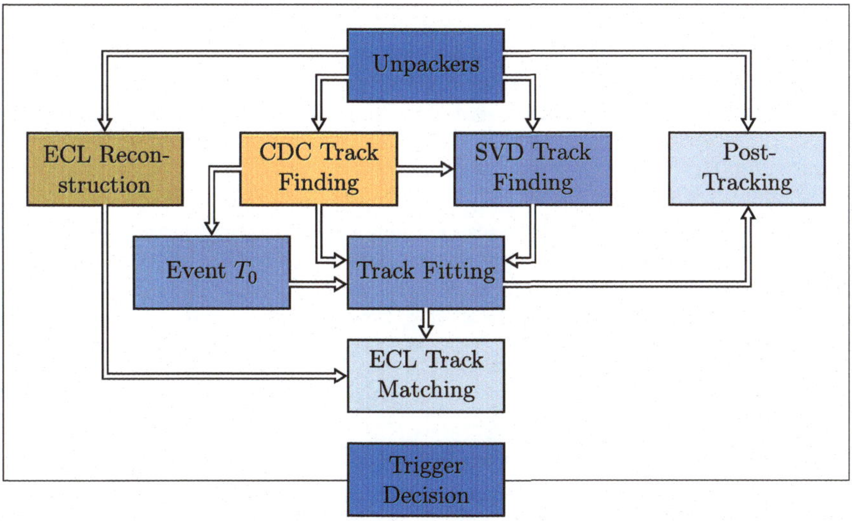

Fig. 4.18 Dependencies among the different reconstruction steps required to find the full trigger decision with calibration tags. The names and colors are comparable to Fig. 4.14. Each arrow depicts the result of a calculation step, which is required for the next part

track finding also includes the CKF developed in this thesis, it needs a preliminary track fit of the CDC tracks. The CDC tracks are additionally used to extract the event time T_0 and are merged and fitted together with the SVD tracks. The fitted tracks are extrapolated into the ECL for the calorimeter-track matching and into the remaining outer detectors (e.g. TOP) for the remaining hit reconstruction. The information of all detectors and reconstruction steps is combined for the HLT trigger decision. After some additional recombination of the final state particles to intermediate resonances like J/Ψ, the events can also be tagged if they are usable for the regular calibration. As it can be seen, only the CDC track finding and the ECL reconstruction have no dependency on other reconstruction steps. It is not possible to make a reasonable selection after the unpackers as otherwise the decision would already be possible at L1 level. The Bhabha reduction needs to run at least after the ECL reconstruction or the CDC track finding.

4.3.2 Bhabha Veto

Bhabha events have a distinct topology compared to the remaining channels, which helps to reduce the rate of these events at an early stage. In case of no or little inelastic interaction with the detector material, the produced e^+e^- pair carries the full energy of $\sqrt{s} = 10.573\,\text{GeV}$ in equal parts in the center of mass system (CMS). The e^+e^- pair can therefore be reconstructed as two high-energetic clusters in the ECL. To account for final-state radiation and inefficiencies in the cluster reconstruction, the minimal required energy of each ECL cluster in the CMS is only $E^* = 3\,\text{GeV}$. Figure 4.19 shows the highest and second-highest reconstructed ECL cluster energy in the event. Due to the limited acceptance of the ECL detector, not all events deposit two high-energetic clusters. In those events, the highest or second-highest energy deposition comes from beam-induced background and is therefore significantly smaller. The

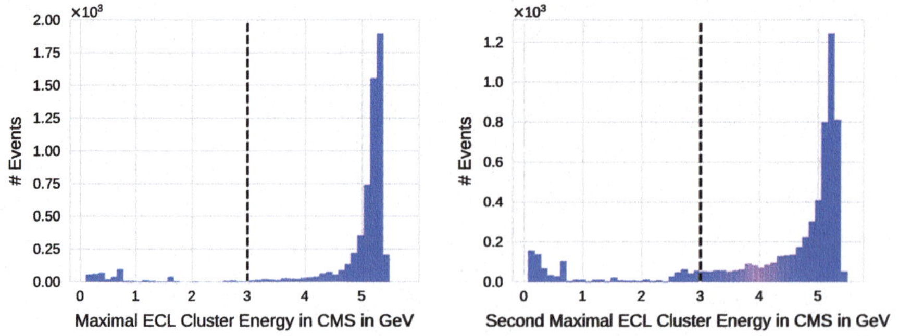

Fig. 4.19 Highest and second-highest reconstructed ECL energy in the CMS for Bhabha events that pass the L1 selection. Due to the detector acceptance, not all events deposit one or two high-energetic clusters in the ECL and the highest measured clusters are from beam-induced background

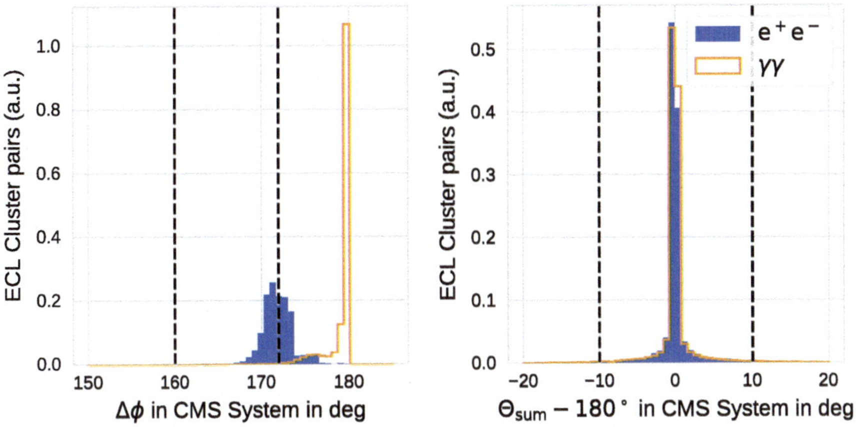

Fig. 4.20 Difference in ϕ^* and sum of θ^* of two high-energetic ECL clusters for e^-e^- and $\gamma\gamma$ events in the CMS. Due to the two-body decay, the two clusters are pointing approximately back-to-back (except for possible final-state radiations), which can be exploited in the Bhabha veto (cuts shown as dashed lines)

requirement of $E^* > 3$ GeV for both clusters dismisses $(23.3 \pm 0.5)\%$ of the correct Bhabha events after the L1 selection.

As the e^+e^- pair originates from a two-body decay, the clusters are separated approximately back-to-back within the uncertainty. Deviations from this can be caused by additional final-state radiation of photons. Figure 4.20 shows the difference in ϕ^* and the sum of θ^* of one ECL cluster pair per event with at least $E^* = 3$ GeV in the CMS system. As expected, the assumption of two back-to-back pairs holds to a high degree. Therefore, cuts of $\Delta\phi^* > 160°$ and $170° < \sum\theta^* < 190°$ will be applied. The cut values were taken from the HLT menu developed for Phase 2 of the experiment [18].

Using these cuts, 71% of the correct Bhabha events after the L1 trigger can be filtered by the early Bhabha veto. Most of the remaining Bhabha events cannot be selected because one or both leptons leave the ECL unobserved outside of the acceptance. Although it would be possible to include cuts on variables based on single clusters, the crucial decision power for the Bhabha selection comes from the back-to-back requirement.

However, Bhabha events are not the only events with this topology in the ECL. Collisions with a high-energetic $\gamma\gamma$ decay can produce a similar topology, which is also shown in the figures discussed above. A large fraction of these physically interesting events would be rejected when vetoing the events only with the cuts described so far. There are two distinct differences between the Bhabha and $\gamma\gamma$ events: In contrast to the photons, the e^+e^- pair is charged and produces one or two distinct tracks in the tracking detector. Also, final state radiation of a γ cannot happen for $\gamma\gamma$ events, so the $\Delta\phi^*$ values are higher on average for these events.

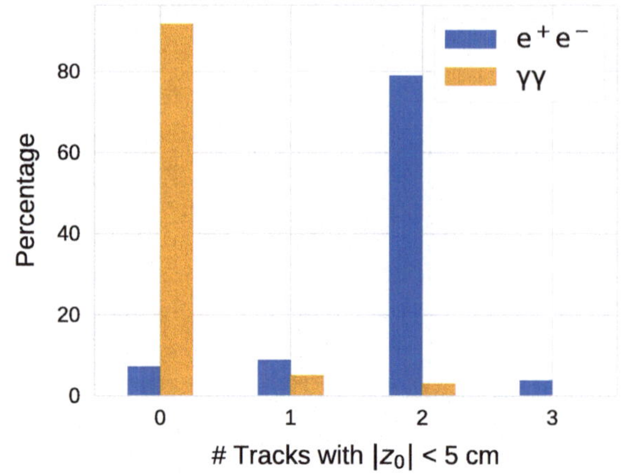

Fig. 4.21 Number of tracks with $|z_0| < 5$ cm found by the CDC track finding algorithm before applying a complex track fit in comparison between e^+e^- and $\gamma\gamma$ events. As the leptons are charged, most of the decay topologies produce at least a single track in the detector

This makes it possible to select Bhabha events with a $\Delta\phi^*$ between 160° and 172° only using the ECL information. To reduce the Bhabha rate even more, two charged tracks originating approximately from the IP can be required for the remaining events to distinguish between $\gamma\gamma$ and e^+e^- topologies. For this, it is sufficient to use only tracks found in the CDC before applying a proper track fit, as the preliminary information after the track finding is good enough for this. Figure 4.21 shows the number of tracks with $|z_0| < 5$ cm, which tags them as approximately originating from the IP. As this cuts away tracks which could have been originating from decays of photons, this variable can be used to distinguish between two-photon and Bhabha events.

The proposed exemplary Bhabha veto developed in this section therefore implies to run the ECL reconstruction with a first cut

$$E^* > 3\,\text{GeV (for two clusters)}, \quad 160^\circ < \Delta\phi^* < 172^\circ, \quad 170^\circ < \sum \theta^* < 190^\circ\,.$$

After that, the CDC reconstruction is performed and a cut is made on

$$E^* > 3\,\text{GeV}, 160^\circ < \Delta\phi^*, 170^\circ < \sum \theta^* < 190^\circ, \text{at least two tracks with } |z_0| < 5\,\text{cm}\,.$$

Fig. 4.22 shows the efficiency of the combined Bhabha veto including the described cut on ECL cluster pairs and the tracks found by the CDC track finder for different event topologies. The first part of the veto for events with $\Delta\phi < 172^\circ$ can be calculated on ECL information only. The second part for the remaining events needs additional CDC tracking. As expected, the efficiency for Bhabha is around 66%, while all other channels are nearly untouched. Especially the important $B\overline{B}$ channels as well as decays with a $\mu^+\mu^-$ or $\tau^+\tau^-$ final state were never dismissed in the large simulation sample. Even the similar $\gamma\gamma$ topologies are remaining in approximately 98% of the cases. In the end, both the information from the ECL reconstruction and the CDC must be used to select a large fraction of Bhabha events without decreasing the efficiency in any other channel. As the efficiencies in this study are small, the used

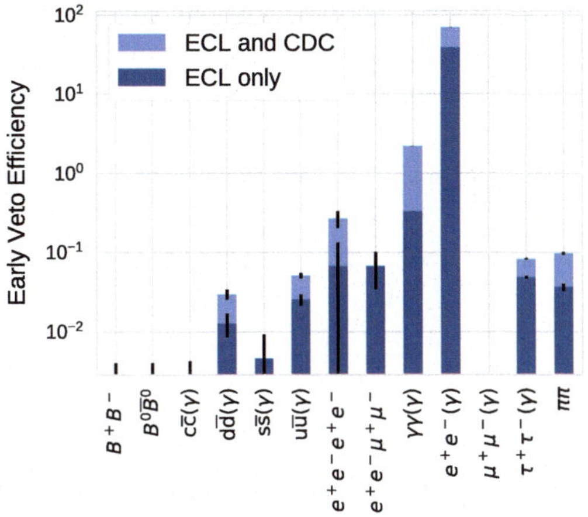

Channel	Efficiency in Percentage
B^+B^-	0.000 ± 0.004
$B^0\overline{B}^0$	0.000 ± 0.004
$u\overline{u}(\gamma)$	0.051 ± 0.004
$d\overline{d}(\gamma$	0.030 ± 0.004
$c\overline{c}(\gamma)$	0.000 ± 0.004
$s\overline{s}(\gamma)$	0.005 ± 0.005
$e^+e^-(\gamma)$	69.318 ± 0.015
$\gamma\gamma(\gamma)$	2.186 ± 0.002
$\mu^+\mu^-(\gamma)$	0.000 ± 0.002
$\tau^+\tau^-(\gamma)$	0.082 ± 0.002
$\pi\pi(\gamma)$	0.096 ± 0.004
$e^+e^-e^+e^-$	0.260 ± 0.070
$e^+e^-\mu^+\mu^-$	0.067 ± 0.033

Fig. 4.22 Efficiencies of the early Bhabha veto described in the text. As expected, especially the e^+e^- channel is vetoed and all other channels including the similar $\gamma\gamma$ remain nearly untouched. The uncertainties were calculated following the description in [24]

bootstrapping method to estimate the statistical uncertainty is not working anymore in these cases. Therefore, the statistical uncertainty was estimated using the formula derived in [24].

The early Bhabha vetos after the first fast-running parts of the reconstruction reduce the number of events, where the full reconstruction needs to run. Figure 4.23 shows the processing time for the most important channels including the early Bhabha veto analogous to Fig. 4.17 weighted by their relative fraction after L1. The processing time for e^-e^+ events is drastically reduced. The average time is now 0.3 s. With this runtime it is feasible to operate the HLT in the planned setup with 20 worker nodes. The processing power is even larger as required making it possible to include additional reconstruction parts if needed. The data rate after the early Bhabha vetos is reduced from the input rate of (21.76 ± 0.27)–(14.23 ± 0.15) kHz which can then further be reduced by the remaining HLT cuts operating on the results of the full reconstruction. The final HLT menu goes beyond the scope of this thesis. It is currently under development and is not in its final form for use in Phase 3 of the experiment.

The implemented Bhabha veto is just one example of an continuous HLT decision and was used to illustrate the feasibility to reduce the reconstruction time by early cut decisions. The main focus has been to reduce the processing time while keeping the efficiency for the interesting physical channels as high as possible without a thorough study of the implications for the physics analyses. As the final HLT and calibration trigger menu is still unknown at this stage, also the early pre-filtering cannot be implemented now in its final version. Additional studies on other event topologies such as other low-multiplicity channels or specific B-decays must be

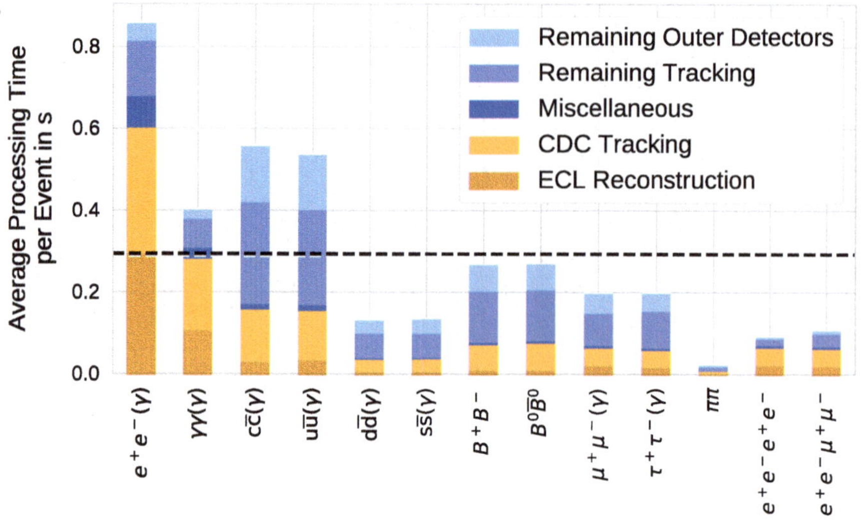

Fig. 4.23 Reconstruction processing time of the different channels analogous to Fig. 4.17, but including an early Bhabha veto. The reconstruction time for Bhabha events is drastically decreased due to the two selections before and after the CDC track finding

performed before the HLT menu can be used in the experiment. The developed framework used for the continuous HLT decision is general enough to be used for further developments and will be described in the following.

4.3.3 General Framework

During the development of the HLT continuous rejection, the framework for the software event trigger was developed from scratch. The currently developed HLT menu is built on top of this framework. With the exemplary fast Bhabha veto, the HLT decision itself and the calibration channel tags, there are multiple different steps in the reconstruction that need to perform customizable cuts and select events. Further developments may add other stages. In the new implementation it is possible to handle all decisions with the same software trigger module and a common software trigger framework. Only a single Bhabha veto will be shown in the following for better visibility.

The basic entity of the software trigger framework are the *software trigger variables*. Each stage in the trigger (e.g. early Bhabha veto, HLT, calibration) defines a set of variables, which are calculated after the reconstruction has reached that stage. Some of the variables needed for the early Bhabha veto were already discussed above. Those variables are only valid for one given stage but can also be stored to the data store or to disk for later analysis.

The calculated variables are used by *software trigger cuts*, which include conditions on these variables concatenated by Boolean operators. The trigger conditions are stored as a syntax tree similar as e.g. the analysis package handles conditions on variables during the event skimming. Additional to the trigger condition an optional prescaling can be stored. Each trigger cut is either of *accept* or *veto* type, which defines its result and its interplay with other cuts in a trigger menu described later. When a cut is checked, the bare cut condition is evaluated and gives either a positive or negative result. The return value of the cut itself is then dependent on the type of the cut and shown in Table 4.3. It can either return `accept`, `reject` or `no result`. The more complex tri-state logic instead of a single Boolean value is needed to also handle veto cuts with the same framework. The prescaling only happens when the return value is `accept`. Multiple prescaling factors can be used to have different factors for different phase space regions. Hereby, the events are classified in bins of the θ parameter of the negative charged track with the highest momentum, which is especially useful for Bhabha events.

The cuts of one trigger stage (e.g. early Bhabha veto, HLT, calibration) are combined into a *software trigger menu*. Each cut in this list is evaluated separately for the final result of this trigger stage. To account for different possible configurations, the trigger menu carries the information what to do in cases of ambiguities. Either the accept or the veto cuts will be preferred depending on the menu type. All possible configurations are discussed in Table 4.3. Despite the more complex tri-state settings of each cut, the final decision of a trigger menu is only a binary flag.

As discussed before, the event reconstruction after each stage is continued only if the event is accepted. If it is rejected, the data store is stripped down to just include

Table 4.3 Top: Return values of the software trigger cut depending on the cut type and the result of the cut condition evaluation. Bottom: Final result for a trigger menu depending on its menu type and the results of the single trigger cuts it contains

Cut type	Cut condition result	Cut result
Veto	Positive	`reject`
	Negative	`no result`
Accept	Positive	`accept` or `reject` depending on the prescaling
	Negative	`no result`

# Cuts with `accept` result	# Cuts with `reject` result	Menu result if	
		Veto overrides	Accept overrides
>0	>0	reject event	accept event
	=0	accept event	accept event
=0	>0	reject event	reject event
	=0	reject event	accept event

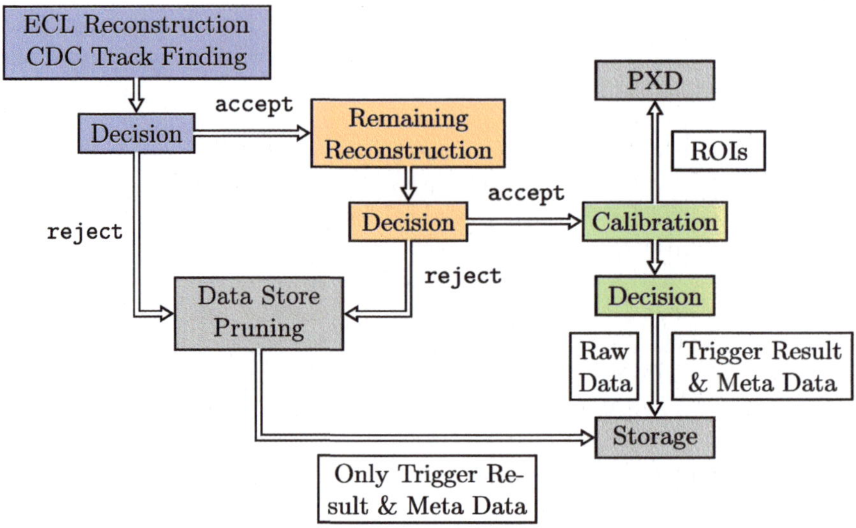

Fig. 4.24 Data flow in the software trigger with the three stages Bhabha veto in blue, remaining HLT reconstruction in orange and calibration in green. Each stage consists of a reconstruction and a decision part. The next stage is only executed in case of a positive decision. For rejected events, only the event meta data together with the trigger decision is stored. Accepted events additionally contain the raw event data and the ROIs, which are only calculated in case of a positive trigger decision. After the calibration, no event selection happens

the basic event information: the trigger decisions made so far and the event meta data like the event and run number. The raw data is kept only if the event is accepted. The full data flow using the exemplary early Bhabha veto is shown in Fig. 4.24. All the single trigger cut decisions as well as the results of each stage are stored into a dedicated object in the data store and can be accessed easily for trigger efficiency studies.

The trigger cut conditions, the prescaling factors, and the trigger menu are stored in the condition database and can be configured for each run of the experiment independently to respond to changing environment conditions, e.g. beam-induced background. Although it is not planned now, the framework can also handle runs dedicated to specific trigger menus, e.g. dark photon searches, which would not be possible otherwise because of the large data rates.

4.4 Multiprocessing Using the ØMQ Library

As already discussed in Chap. 2 and also seen in the processing time studies in this chapter, multiprocessing plays a crucial role for the HLT. The amount of data can only be processed in time when it is distributed reliable and fast to all CPUs of the workers. Additionally, the distribution should be smart, so that all processors are equally charged with work to perform. Those conditions are well fulfilled by the current multiprocessing implementation based on ring buffers.

However, there are also other requirements to the multiprocessing framework in basf2. As it plays such a crucial role for online as well as offline[2] reconstruction, there is the need to create a maintainable framework. Each component needs to have long-term support, as this core functionality will be present until the end of the experiment. Recorded data triggered by the L1 trigger that is somehow lost during the HLT reconstruction cannot be recovered. This means, the reconstruction needs to be safe against crashes of the software or at least not lose data in these cases. Additionally, stuck or broken worker processes must be stopped and replaced to restore the full processing power of the HLT nodes. For offline reconstruction during development and especially for simulating MC events, the event size passed between the processes can be much higher compared to the recorded data, as intermediate reconstruction objects or additional MC truth information must be transferred. The framework needs to cope with these larger messages and support basically unlimited event sizes.[3] Finally, it is beneficial to handle the multi-core as well as the multi-node calculation within the same framework which should be open for new developments as high-performance computing (HPC) or high-throughput calculation (HTC).

Unfortunately, those additional requirements are not fulfilled by the current multiprocessing implementation based on ring buffers. Although built on the widely used technology of ring buffers, the framework uses mostly code written only for basf2 without any further users outside of the collaboration, which decreases the number of test cases. It does not allow for any communication between the processes except the event data, which makes it impossible to include features as automatic restart of failed processes or backup of crashed workers. The ring buffer size is fixed at the start of the calculation and currently cannot handle larger simulated events with a lot of MC data. Additionally, the technique of ring buffers can only be used for multi-core calculations.

This section proposes a new multiprocessing framework for basf2 based on the open-source ØMQ (pronounced ZeroMQ) library [25]. The new framework builds upon the message passing design pattern [26] and fulfills all the requirements discussed before. It was designed and implemented during this theses, together with Thomas Hauth and Anselm Baur [27], and successfully tested on parts of the HLT hardware.

---

[2]The grid sites used for simulating as well as reconstructing events after the recording will allow for jobs using multiple cores in the future. Also the local development today is mostly performed on multi-core machines where multiprocessing can speed up the development time significantly.

[3]As this is of course not possible because of hardware constraints, the limit should at least be much larger than the average raw data size.

4.4.1 Overview on ØMQ

ØMQ, as described by its authors, "looks like an embeddable networking library but acts like a concurrency framework. It gives you sockets that carry atomic messages across various transports like in-process, inter-process, TCP, and multicast. You can connect sockets N-to-N with patterns like fan-out, pub-sub, task distribution, and request-reply. It's fast enough to be the fabric for clustered products. Its asynchronous I/O model gives you scalable multi-core applications, built as asynchronous message-processing tasks" [25].

The quote already gives a good overview of the benefits of the ØMQ library for the use as a multiprocessing framework. More information on the technical details of the library can be found in the rest of the manual [28]. Tests performed by other collaborations have already shown, that the library is usable in high energy particle physics applications [29, 30].

The basic ingredients in the ØMQ library are generalized sockets, which implement an asynchronous message queue via an additional non-blocking background thread. Each particular socket type is optimized to support a distinct message passing pattern, e.g. request-reply or publish-subscribe, and can be used for 1-N or N-N communications. The library handles the message passing via those sockets automatically. The user can hereby choose which communication channel—in-thread (ITC), inter-process (IPC) or transmission control protocol (TCP)—is used for a given socket making the step from multi-core to multi-node calculations simple. This thesis uses the IPC transport mode.

The ØMQ library is licensed with LGPL and developed in a large community. It is used for many projects in different branches, as there exist many third-party bindings for different popular programming languages (e.g. Python and C++ [31]).

4.4.2 Implementation for basf2

The new multiprocessing implementation employs the message passing design pattern for concurrent processing. The implementation was first described in [27] and is repeated here briefly. The input, output and worker processes only communicate with messages sent using the ØMQ library. The event data is still transferred from the input via the worker to the output process by direct communication. Additional event messages such as termination signals or event backups can also be sent via a message proxy between each forked process or to their parent process, which operates as a monitoring instance. Figure 4.25 shows a basic overview on the process types and their communication in the new implementation.

Each module in basf2 is tagged with a flag by the developer, if it is able to run in multiprocessing mode not needing any singleton resources. Analogous to the current multiprocessing implementation, the framework starts by splitting up the module path into three parts based on this flag: the input part with the largest coherent piece of

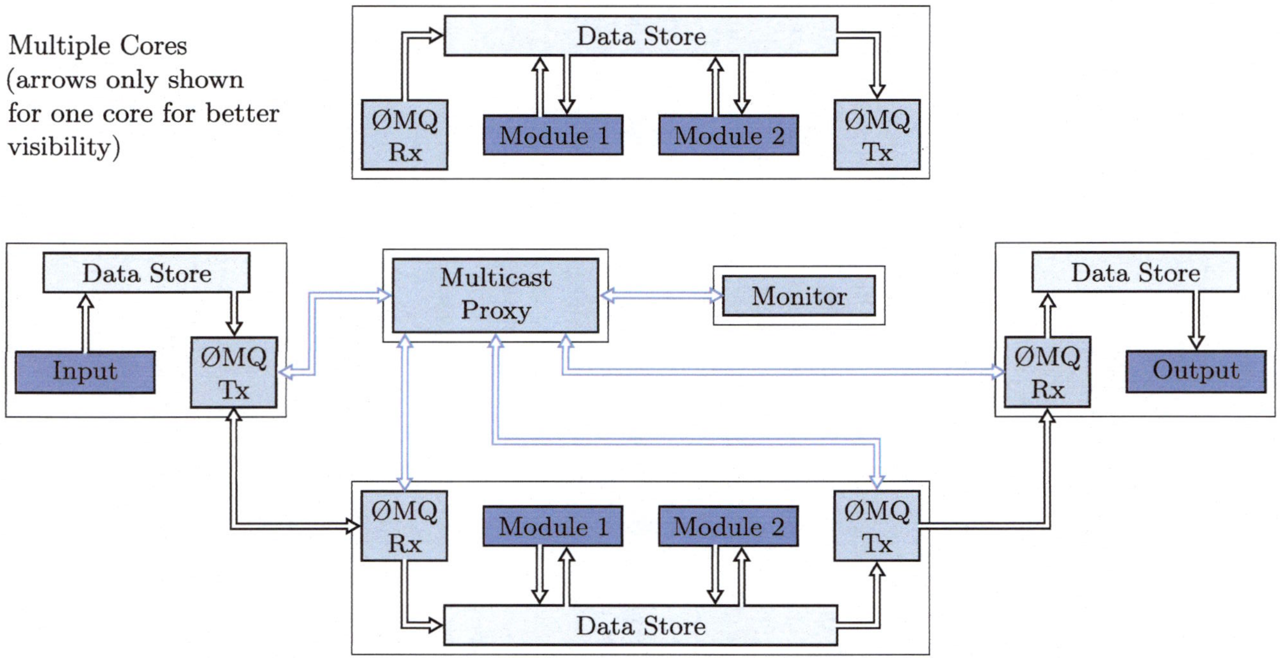

Fig. 4.25 Basic building blocks of the new multiprocessing framework based on the ØMQ library comparable to Fig. 2.8. The input, output and worker are again implemented as distinct processes which all process a distinct part of the basf2 module path. The new implementation has an additional monitoring and message proxy process. As it is the case in the current implementation, special modules in the beginning or end of the paths handle the event and message transferal. The event data transferal (black arrows) happens via direct socket communication whereas event messages are distributed via the multicasting proxy (blue arrows)

non-multiprocessing modules, the worker part with all multiprocessing modules and the output part with the remaining non-multiprocessing modules. The different parts of the path are processed by their corresponding processes until either the end of the data stream, an external interrupt by the user or an error occurred. ØMQ transmission (Tx) and receive (Rx) modules are included at the beginning and end of the path pieces handling the data communication between the processes. The task of the additional proxy process is to transport incoming event messages to all listening subscribers. By using a distinct process for this, all the publishers and subscribers only need to connect to a single stable endpoint, which is beneficial in cases of restarted processes. As the messages sent via multicasting are rather small (see [27]) and rare, the proxy is mostly idling and therefore not degrading the overall processing power of the machine. The monitoring process is the parent of all spawned processes. Its task is to supervise the data and event message flow and its child processes and to terminate or spawn additional processes if needed.

To understand how the different building blocks play together, the data flow for a single event will be discussed in the following. Before forking, all modules in the path are initialized in the parent process. Due to the implementation of the fork system call and the copy-on-write technique used in modern operation systems [32], the loaded data and libraries during implementation can be used by all child processes without the need for additional memory allocations. After forking, each process is at first sending a *hello* message via the multicast. Due to this, the monitor knows which processes have come alive and the input and output processes know, how many worker processes are there. As this may change during runtime, this information is spread on every new start of a process. When the worker processes have started, their Rx modules send as many *ready* messages via the direct connection to the Tx modules of the input process as their event buffer size can hold.[4] For every *ready* message it receives, the Tx module of the input process streams the data of a single event via the direct connection back. When the event data is received, the Rx module of the worker process returns another *ready* message. This principle of data distribution assures that processes working on events, which need a longer processing time do not get an event as fast as other processes, thus implying a basic balancing between the workers. After the event is processed by the different modules in the worker path, the Tx module of the worker sends the event to the Rx module of the output process. The event loop in the worker then jumps back to its Rx module, which waits until it receives the data for the next event from the input process. As ØMQ uses a buffered message sending and the workers sent multiple *ready* messages, chances are high that the input process has already sent a new event while the worker process was still processing the old one. This decreases the waiting time with the cost of slightly higher memory consumption because multiple events have to be kept in the buffers. As the event size at Belle II is small, this is not a problem in practice. The Rx module of the output receives the event from the worker process and stores it according to the output modules in its path.

[4]The event buffer size can be controlled via module parameters and set to two by default.

The event data and the ready messages are however not the only communication happening during the processing. Additional messages are sent for acknowledgement and safety. More information on these advanced features such as the event backup can be found in [27].

The modular design together with the lock-free message passing implementation based on the open-source library ØMQ make the new algorithm fast and easily maintainable. The additional multicast makes the handling of crashing or hanging processes possible. As the message size is not limited in ØMQ, it is possible to also handle MC simulations with a large amount of truth information. Also, the setup can be adapted to be used for multi-node calculations using network communication by just changing the message transport from IPC to TCP. Even applications of other protocols typical in HPC or HTC are possible [30].

4.4.3 Performance

The new multiprocessing framework based on the ØMQ library has clear benefits compared to the old ring buffer in terms of features, extensibility and maintainability. It is however clear that although these benefits are crucially needed, they should not come with the cost of additional processing time. For this, the important parts of the processing time studies presented earlier in this chapter are repeated with the new implementation. Some results of this can also be found in [27]. Figure 4.26 shows the difference in the processing time per event between the current and new framework. The calculation was performed with the same procedure as discussed in the previous studies: the input data was stored in the raw format on a RAM disk to allow for fast access and the logging was turned off. Both large B^+B^- as well as smaller e^-e^+ events were tested. No significant differences could be measured between the two implementations. Albeit the new implementation has many additional features, it does not slow down the reconstruction process measurably. This also implies that the scaling behavior is similar, which is expected as the cache misses influencing the old implementation are of course still present in the new implementation.

To test the new features of the ØMQ implementation, artificial crashes were introduced into the event processing. According to a specific crash rate, distinct events were chosen where a kill signal was send to one of the workers. A random jitter was applied to the event numbers to simulate a real reconstruction crash. To have a controlled environment, the remaining reconstruction was replaced by modules waiting a specific amount of time instead of a full reconstruction. In Fig. 4.27 the number or running worker processes at each time step during the reconstruction is shown. The reconstruction was performed with four worker cores in parallel. The process count is measured by periodically accessing the number of child processes using tools of the linux operation system [33]. The measurement frequency was limited to $f = 1\,\text{kHz}$ to not overload the machine due to the measurement.

The introduced artificial crashes are clearly visible as a drop in the number of processes. In all cases, the framework was able to recover from the crash by restarting

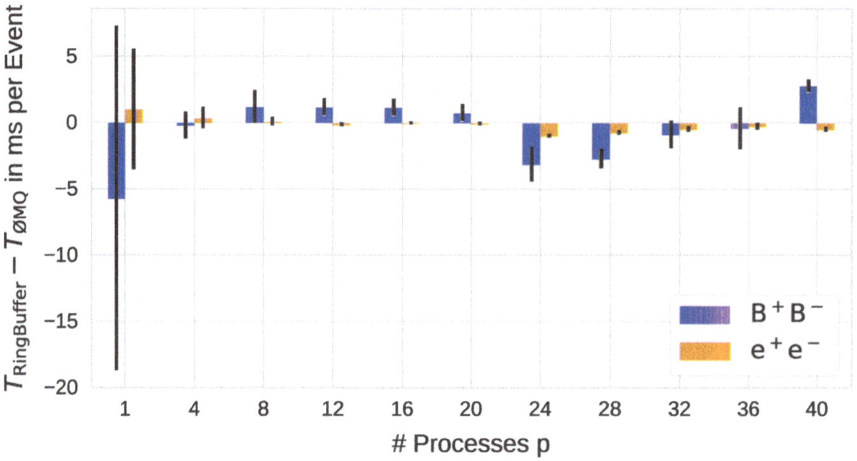

Fig. 4.26 Difference in the measured processing times per event between the new and old implementation of the multiprocessing shown for B^+B^- and e^-e^+ events. Although the new implementation has additional features, no measurable difference between the reconstruction times is visible. In this study, the full processing time including the burn-in and termination phase is used to also account for possible additional initialization time penalties

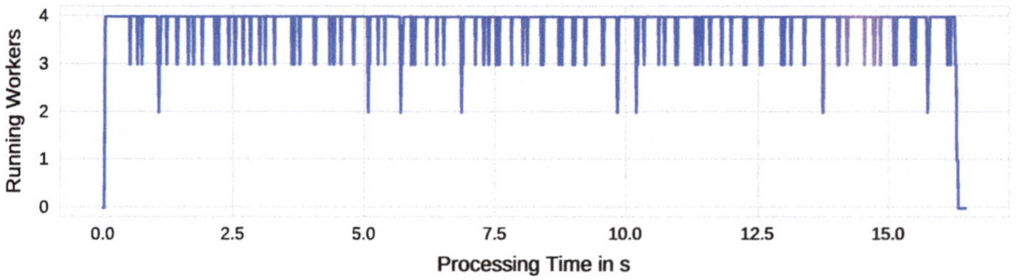

Fig. 4.27 Measured number of running worker processes during the reconstruction of an artificial path, where periodically chosen events will crash the workers by sending a kill signal to the process. The new ØMQ implementation is able to recover all failed workers within a single measurement period

the worker process and continuing the event processing within the next iteration of the process count measurements. With the measurement frequency of $f = 1$ kHz the maximal time needed to restart a single worker is therefore $T = 1$ ms, which is fast enough for the usage on the HLT.

If a worker process dies or is killed because of a too large processing time, the event data send to this worker will never reach the output process via the normal data flow. Therefore, the events are send directly to the output and flagged as not reconstructed properly. Figure 4.28 shows the fraction of events with this flag depending on the fraction of events, where one worker was killed artificially. For each killed worker, the current event where the crash happened, as well as all events in the event buffer of the worker, need to be send directly to the output. As the default size of the event

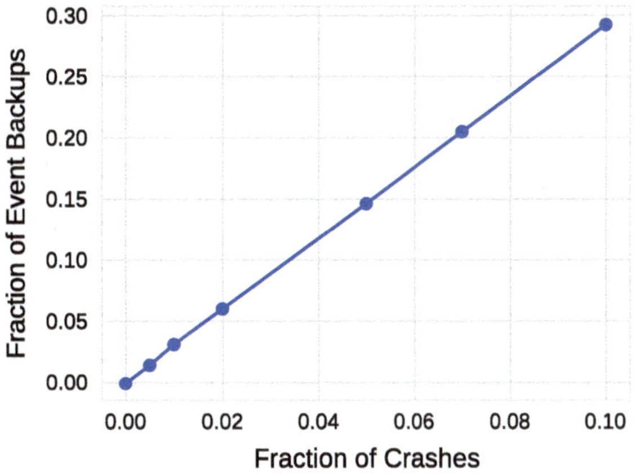

Fig. 4.28 Fraction of events which are sent directly to the output process without reconstruction (backup) due to failed worker processes for artificially introduced worker crashes. As not only the current event but also all events in the event buffer of the worker must be backed up, the fraction of backup events is three times larger

buffer is two, for each killed worker three events are sent directly to the output, which is also visible in the figure. The anticipated fraction of killed workers due to timeouts or crashes is much lower than presented in this study. The number of events without reconstruction and therefore also without HLT tags can then easily be handled manually offline.

4.5 Summary

The HLT reconstruction plays a crucial role for the success of an experiment. In this work, it was first shown that it is possible to reconstruct the events in time using a specialized version of the offline reconstruction and a continuous HLT trigger decision. During this, the software for the HLT trigger decisions was written. As an example, an early Bhabha veto was first developed here. Additionally, the multiprocessing speedup of the software was studied in detail and possible scaling issues were examined. As a replacement for the legacy multiprocessing framework based on ring buffers, a new version with the ØMQ library was introduced. The new framework solves many problematic issues of the old implementation and was successfully tested in this thesis.

References

1. Urquijo P, Ferber T (2016) Overview of the Belle II physics generators. https://docs.belle2.org/record/282/files/BELLE2-NOTE-PH-2015-006-v2.pdf
2. Iwasaki Y, Cheon B, Won E, Gao X et al (2011) Level 1 trigger system for the Belle II experiment. IEEE Trans Nucl Sci 58(4):1807–1815. http://ieeexplore.ieee.org/document/5740389/, https://doi.org/10.1109/TNS.2011.2119329

3. Shiu J-G (2016) The level 1 trigger system for Belle II CDC. In: 2016 IEEE-NPSS Real Time Conference (RT), pp 1–3. IEEE. http://ieeexplore.ieee.org/document/7543137/
4. Kim S, Lee I, Unno Y, Cheon B (2017) Status of the electromagnetic calorimeter trigger system at Belle II. J Phys Conf Ser 928:012022. http://iopscience.iop.org/article/10.1088/1742-6596/928/1/012022/pdf
5. Belle II Collaboration (2019) Basf2 software repository trigger package. https://stash.desy.de/projects/B2/repos/software/browse/trg?at=refs%2Ftags%2Frelease-02-00-01
6. Kuhr T, Pulvermacher C, Ritter M, Hauth T, Braun N (2018) The Belle II core software. arXiv:1809.04299
7. Intel Corporation (2014) Intel® Xeon® Processor E5-2660 v3 (25M Cache, 2.60 GHz) Produktspezifikationen. https://ark.intel.com/de/products/81706/Intel-Xeon-Processor-E5-2660-v3-25M-Cache-2_60-GHz
8. Amdahl GM (1967) Validity of the single processor approach to achieving large scale computing capabilities. In: Proceedings of the April 18–20, 1967, Spring Joint Computer Conference on–AFIPS '67 (Spring). ACM Press, New York, USA. http://delivery.acm.org/10.1145/1470000/1465560/p483-amdahl.pdf?ip=129.13.102.208&id=1465560&acc=ACTIVESERVICE&key=2BA2C432AB83DA15.50F04DEC6D8CC69A.4D4702B0C3E38B35.4D4702B0C3E38B35&__acm__=1537457854_f3487ed1102f3a71b6429db4d2d5670a, https://doi.org/10.1145/1465482.1465560
9. Yamada S, Itoh R, Nakamura K, Nakao M et al (2015) Data acquisition system for the Belle II experiment. IEEE Trans Nucl Sci 62(3):1175–1180. http://ieeexplore.ieee.org/document/7117478/, https://doi.org/10.1109/TNS.2015.2424717
10. Asner DM, Dart E, Hara T (2013) Belle II experiment network and computing. arXiv:1308.0672
11. Marr DT, Hinton G, Koufaty DA, Miller JA (2002) Hyper-threading technology architecture and microarchitecture. http://www.cs.virginia.edu/~mc2zk/cs451/vol6iss1_art01.pdf
12. Intel Developer Forum (2002) Intel demonstrates breakthrough processor design. https://www.intel.com/pressroom/archive/releases/2001/20010828comp.htm
13. Levinthal D (2008) Performance analysis guide performance analysis guide for Intel® Core™ i7 processor and Intel® Xeon™ 5500 processors. http://www.intel.com/technology/hyperthread/index.htm;
14. Vitillo RA (2011) Performance tools development. Futur Comput Part Phys 16. https://indico.cern.ch/event/141309/contributions/1369454/attachments/126021/178987/RobertoVitillo_FutureTech_EDI.pdf
15. Salzburger A (2015) Optimisation of the ATLAS track reconstruction software for run-2. J Phys Conf Ser 664:072042. IOP Publishing. http://stacks.iop.org/1742-6596/664/i=7/a=072042?key=crossref.c1d153f8aa8896687664b45b57a803d5, https://doi.org/10.1088/1742-6596/664/7/072042
16. Meyer J (2016) Optimisation of the material parameters for track fitting in the central drift chamber of the Belle II experiment. Bachelor's thesis, KIT. https://ekp-invenio.physik.uni-karlsruhe.de/record/48884/files/EKP-2017-00011.pdf
17. Gumpert C, Salzburger A, Kiehn M, Hrdinka J, Calace N (2017) ACTS: from ATLAS software towards a common track reconstruction software. J Phys Conf Ser 898:042011. http://iopscience.iop.org/article/10.1088/1742-6596/898/4/042011/pdf, https://doi.org/10.1088/1742-6596/898/4/042011
18. Hearty C (2018) Jira issue on HLT phase 2 menu. https://agira.desy.de/browse/BII-3617?focusedCommentId=44831&page=com.atlassian.jira.plugin.system.issuetabpanels%3Acomment-tabpanel#comment-44831
19. Adam W, Bergauer T, Deldicque C, Erö J et al (2006) The CMS high level trigger. Eur Phys J C 46(3):605–667. http://www.springerlink.com/index/10.1140/epjc/s2006-02495-8
20. Nakamura T (2001) Study of the level 3 software trigger system with the silicon vertex detector for the belle experiment. Master's thesis, Tokyo Institute of Technology. http://www.hp.phys.titech.ac.jp/dmthesis/Mnakamura.pdf

21. Li C (2016) Study on track and ECLCluster reconstruction in the HLT. In: Belle II General Meeting. Tsukuba. https://kds.kek.jp/indico/event/21740/session/21/contribution/187/material/slides/0.pdf
22. Braun N, Hauth T, Li C (2016) Studies on level 3 and HLT performance. In: Belle II General Meeting. Tsukuba. https://kds.kek.jp/indico/event/21740/session/21/contribution/186
23. Ferber T, Hearty C, Roney JM (2016) Design of the ECL software for Belle II (version 2.0). https://docs.belle2.org/record/316/files/BELLE2-NOTE-TE-2016-001.pdf
24. Ullrich T, Xu Z (2007) Treatment of errors in efficiency calculations. http://th-www.if.uj.edu.pl/~erichter/dydaktyka/Dydaktyka2011/LAB-2011/0701199v1.pdf
25. Hintjens P (2014) ØMQ–The Guide–ØMQ–The Guide. http://zguide.zeromq.org/page:all
26. Hohpe G, Woolf B (2004) Enterprise integration patterns: designing, building, and deploying messaging solutions. Addison-Wesley, Boston
27. Baur A (2018) ØMQ implementation for a reliable parallel processing framework on the high level trigger at the Belle II experiment. Bachelor's thesis, KIT. https://ekp-invenio.physik.uni-karlsruhe.de/record/49048/files/EKP-2018-00027.pdf
28. iMatix Corporation (2012) ZeroMQ API–0MQ Api. http://api.zeromq.org/
29. Dworak A, Ehm F, Sliwinski W, Sobczak M (2011) Middleware trends and market leaders 2011. http://zeromq.wdfiles.com/local--files/intro%3Aread-the-manual/MiddlewareTrendsandMarketLeaders2011.pdf
30. Rybalchenko A (2015) FairMQ data transport for online & offline processing. In: ALICE Offline Week. https://indico.cern.ch/event/400521/contributions/1843043/attachments/802159/1099377/offline.pdf
31. iMatix Corporation (2014) ØMQ language bindings–zeromq. http://zeromq.org/bindings:_start
32. Wolf J, Wolf K-J (2016) Linux-Unix-Programmierung: Das umfassende Handbuch, 4th edn. Rheinwerk Computing, Bonn
33. Rodola G (2018) psutil documentation—psutil 5.4.8 documentation. https://psutil.readthedocs.io/en/latest/

Chapter 5
Event Timing

The event time describes the time the collision producing a physically interesting event with a positive trigger decision happened. Because of the integration time of the subdetectors, each recorded event includes the data of more than a single collision. However, as the cross sections for the physics channels under study is low compared to the uninteresting background processes (see Chap. 4), the probability for more than one physically interesting topology in a single event is vanishing. If the reconstructed event time is precise enough to decide in which bunch-on-bunch crossing the event happened, the very precise accelerator revolution frequency can be used to refine it even further. Most of the time, the event time is measured as a time difference, e.g. with respect to the L1 trigger time.

The event time plays a crucial role at different stages of the reconstruction. Even before any software reconstruction can run, the data needs to be read out from the detector front end boards. As described in Sect. 2.2, this is handled by the general decision logic, which requests the subdetectors to deliver the data for a given event time. This event time is estimated on the Level 1 (L1) trigger using a subset of hits in the CDC and the ECL clusters with the highest energy. The different subdetectors then decide according to this event time which parts of their storage buffers are sent to the event builder. If the jitter of the trigger time (the deviation from the actual value) is too large, it can happen that the returned data does not include all data related to the topology under study. Especially the SVD has very hard requirements on the precision of the event time [1], as its readout time window cannot be large because of bandwidth constraints. The time extraction on the L1 hardware trigger is not part of this thesis and is described elsewhere [2].

N. Braun, *Combinatorial Kalman Filter and High Level Trigger Reconstruction for the Belle II Experiment*, Springer Theses,
https://doi.org/10.1007/978-3-030-24997-7_5

5.1 The Role of the Event Time

Even if one assumes an ideal data readout with the full event data despite a possibly wrong trigger time, the event time still heavily influences the online and offline reconstruction performed by the software. Most of the detectors (e.g. the TOP, the CDC or the ECL) measure the time of the incoming signals. Since the trigger time may differ from the correct collision time value, this jitter must be compensated during the software reconstruction for extracting the correct information from these subdetectors. Additionally, if the extracted event time in software has a better resolution than the trigger event time, large fractions of the off-time background that are still in the readout window of the subdetectors can be rejected. Especially the background suppression in the ECL benefits largely from a more precise event time. The CDC needs a proper event time to convert the measured ionization arrival time at the wires into a drift length (cf. Fig. 2.4). The drift lengths are used during track finding and fitting. Fortunately, it is possible to start track finding with approximate values for the event time and already end up with a usable result. However, for refinement and for the final track fit, a correct estimation of the event time is necessary.

A bunch crossing occurs approximately every 4 ns and the accelerator clock signal can be used to fix the time, once the correct bunch crossing is known. The goal of the time extraction running on the HLT and in the offline reconstruction is therefore to always be more precise than 2 ns to select the correct bunch crossing. Hereby, the final time extraction can only be performed by the TOP detector, which reaches time resolutions in the ps range. However, for this to work, a proper time hypothesis has to be formulated in advance [3]. The goal of this preliminary extraction is not to reach the precision of 2 ns already, but to be as close as possible to this requirement. Additionally, as the TOP reconstruction needs to run at a later stage of the software reconstruction, other earlier reconstruction steps can benefit from those preliminary event time hypotheses.

In the following, T_0 is defined as the global time the collision event under study happened with respect to the L1 trigger time. T_0 will also be called event time in the following. It is the task of the software reconstruction to extract T_0. The value of the event time T_0 corresponds to the jitter of the trigger time and the two terms will be used interchangeable from now on. If the L1 trigger time was perfect, the extracted event time T_0 would be zero. The trigger jitter is expected to be in the order of 30 ns although the exact distribution is still unknown and depends on the event topology. Therefore, the presented studies will cover the whole spectra of trigger jitters between -60 and 60 ns. No assumptions about the form of this distribution will be made, except that the trigger jitter is quantized with the bunch crossing time of approximately 4 ns.

The following chapter describes the technique for extracting T_0 using the CDC hit and track information, which is already able to achieve a time resolution better than what was initially thought to be possible. Section 5.2 gives an overview on the implemented methods and Sects. 5.3–5.6 explain the methods in detail with their performance. Section 5.7 completes with a discussion on the final results of all methods and their interplay with other time extraction implementations at Belle II, e.g. using the ECL or the TOP. The methods in Sect. 5.6 were developed by Hauth and Braun [4] whereas the rest of the work described in this chapter was performed by the author of this thesis.

5.2 Event Timing Using the CDC Information

As described in Sect. 2.1.2, the measured time for each CDC hit has multiple ingredients:

$$T_{\text{meas}} = T_0 + T_{\text{flight}} + T_{\text{drift}} + T_{\text{prop}} \,, \tag{5.1}$$

where the term for the time walk was already neglected. Figure 5.1 shows the distribution of the measured time information of all CDC hits accumulated over multiple events with an event time of -20 ns to simulate the trigger jitter. As all other contributions to T_{meas} are positive, the measured time of signal hits can never be smaller than the event time T_0. The trigger jitter shifts the signal distribution in time as the trigger time is the reference for the readout—this is the reason why the readout window of the CDC starts already at -100 ns. The plot also shows the distribution for hits produced in background processes. The distribution of these background processes is flat in time, but as wires once hit by an ionizing particle cannot record a second time during this readout window, there is a strong tendency towards smaller drift times. This effect, however, is overestimated in the simulation leading to a sharper rise in the number of background hits for small drift times and lengths. A correct simulation is not possible at this stage and the only possibility for a proper model is to use random triggered recorded data, which is not available for the Phase 3 setup yet.

If the trajectory of the particle is already known, the contributions of T_{flight}, T_{drift} and T_{prop} can be calculated without knowing the value of the event time. The formula shows, that by comparing the measured time value with the sum of the contributions for each track, the event time can be extracted. This is the basis for the methods implemented in this thesis and described in detail in the following.

The general procedure for the time extraction in the CDC is:

1. Do track finding in the CDC detector. This produces sets of hits which belong to the same track. During this process, the drift length and also the event time have to be known. However, the track finding is rather stable against wrong assumptions, so using the event time estimation from the trigger instead of the correct event time is sufficiently accurate (see Sect. 5.3.1).
2. For each track, estimate the flight time T_{flight} from the interaction point until the hit is reached. An input for this is the flight length between the collision point and the current hit, which can be calculated using the track parameters. The

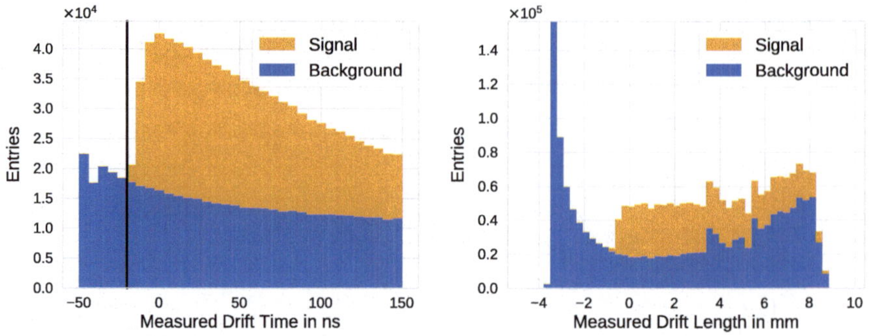

Fig. 5.1 Distribution of the time signals and derived drift lengths measured in the CDC in 100 generic $B\overline{B}$ events with beam-induced background. To simulate the trigger jitter, a fixed $T_0 = -20$ ns was selected. For reconstruction, an assumption of $T_0' = 0$ ns is used. As the assumed time T_0' is 20 ns away from the correct event time T_0, some drift length calculations lead to an unphysical negative drift length. For the background hits there appears a strong peak at the lower end of the time window, which is caused by wrongly simulated background processes

track information also fixes the propagation time T_{prop}—although an additional calibration on data is advantageous (see e.g. [5]).

3. Under the assumption that all tracks in an event come from the same collision,[1] they all share the same value for the event time. A global optimization of T_0 between the measured times and the ones calculated with the right side of Eq. (5.1) leads to the best estimation for the event time. Different algorithms are available for this task and are described in the following sections. Algorithms that run in the online reconstruction like the final track fit or the TOP reconstruction use this new event time hypothesis.
4. In principle, the new refined event time hypothesis can be used to rerun the full reconstruction including track finding and track fitting. However it can be shown, that the performance of the online software trigger decision does not rely strongly on the precise knowledge of the event time [6]. The reason is, that the efficiency of finding high-p_T tracks, which is the most valuable input to the trigger decision from the track reconstruction, is rather stable against wrong time hypotheses.
5. The reconstruction is repeated offline with the extracted information on the event time and refined calibration and alignment constants.

[1] Since the probability to have multiple collisions with tracks is low for a detector readout, this assumption is valid.

5.3 Flight Time Estimation

Estimating the flight time of the particles from the interaction point to each hit is only possible in case the tracking parameters are known well enough. The track parameters can either be taken from the track parameters seeds calculated using a fast χ^2-based fit in the track finding step or a more advanced DAF in the track fit. However, as the correct event time is not known at this stage, it is not obvious if the track finding or fitting still gives reasonable results in these cases.

5.3.1 Performance of the Track Finding with Wrong T_0 Assumptions

As described in Sect. 3.2, the CDC track finding algorithm uses the drift length information during its procedure. With a large trigger jitter, this information can deviate from the correct length by a large value. Figure 5.2 shows the performance of the full CDC track finding algorithm for different simulated event times. It is clearly visible, that the finding efficiency drops by a large fraction for strong deviations from the correct time value indicating once again the need for a correct time extraction. However, the goal in this first iteration of the track finding with a possibly wrong time estimation is not to find all tracks and understand the whole event correctly, but just to extract as much information as needed to estimate the correct time and to come to a trigger decision. This extraction can be performed on a smaller set of tracks in each event. In the following, only tracks with a minimum transverse momentum of 350 MeV and at least 20 hits in the CDC will be taken into account. This selection minimizes the problems caused by secondary decays or wrong particle hypothesis as will be shown below. Figure 5.2 also shows the number of MC tracks that fall into this category in each event as well as the found tracks in this category. It can be seen, that the number of found tracks is sufficiently high also for the shown large deviation of −60 ns. To decrease the processing time, the track set used for the following track studies will be limited to four tracks per event.

5.3.2 Track Time Estimation

A first step in the flight time estimation is to calculate the time the particle needs from the collision point to the first hit. For all following hits, the flight time is estimated using this first estimation—the track time—during the extraction in the Kalman filter from one hit to the next. The current velocity is known from the track state.

For estimating the track time, different algorithms are implemented—depending on which detector setup is used.

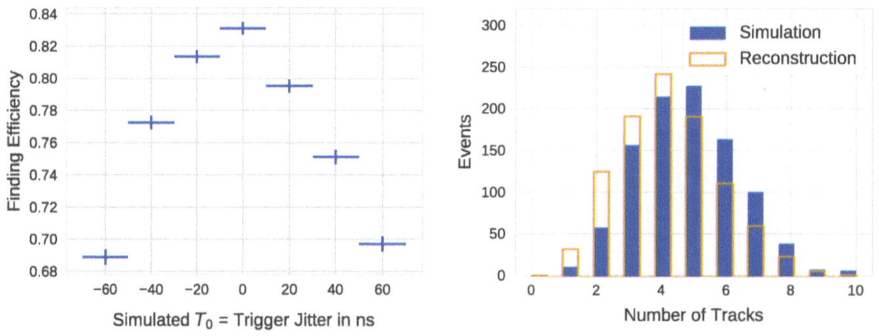

Fig. 5.2 Efficiencies of the CDC standalone track finding calculated on a sample of generic BB events with beam induced background for different assumed event times T_0 using only primary particles (left). In each event, a track selection using only tracks with at least 350 MeV transversal momentum and at least 20 CDC hits is performed and the number of remaining MC tracks and found tracks (right) is shown for an exemplary T_0 value of −60 ns. Although the trigger jitter is so high, the tracking is still able to find most of the selected tracks

Plane Trigger Track Time Estimation During the cosmic run tests performed before the full detector was installed in place, a scintillator plane in the detector center was used for triggering events, as the full trigger setup needed to be tested with an external input for comparison. Instead of the GDL requesting the data for a certain T_0, the external trigger plane gave the time input. In this setup, the track seed or fit state needs to be extrapolated to the plane position. The extrapolated distance together with the muon mass hypothesis leads to the track time. Using just the muon hypothesis is good enough for cosmic ray events.

IP Track Time Estimation In events recorded in Phase 2 or Phase 3 the GDL determines the event time under the assumption that the tracks emerge from the interaction point—although this assumption may only hold approximately in case of cosmic rays. Again, the seed parameters or the fit result is extrapolated to the interaction point. This time, a pion mass hypothesis is used, as the correct hypothesis is unknown at this stage.[2]

The methods described previously to estimate the flight time for each hit assumes that the particles travel without delay from the primary collision to the position of the hit wire. This assumption does not hold anymore, when the particles come from a secondary decay or a decay in flight or when the deposited energy is produced by material effects, e.g. δ-electrons. The influence of δ-electrons will be discussed further down and is only a small deviation. The influence of a wrong hypothesis can be seen in Fig. 5.3a where the estimated flight time using a pion hypothesis is compared to the simulated flight time for different particle types. It can be clearly seen, that the flight time estimation works with a high precision when using the

[2]Assuming a hit in the first CDC layer, the difference between the pion and the proton mass hypothesis is only in the order of 0.2 ns. Additionally, the momentum requirements discussed below cut away protons in most of the cases.

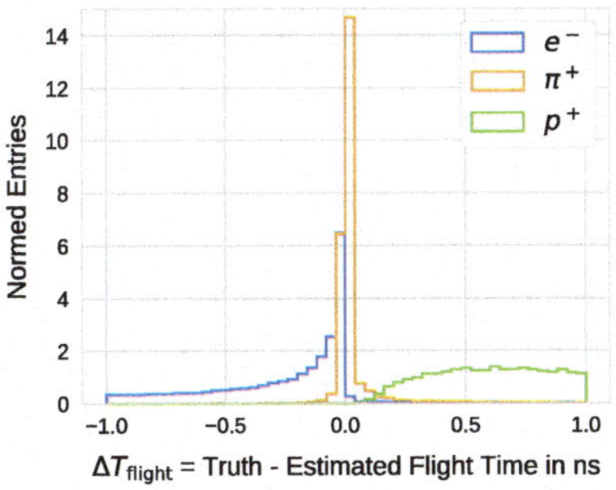

(a) Residual ΔT for different simulated particle types.

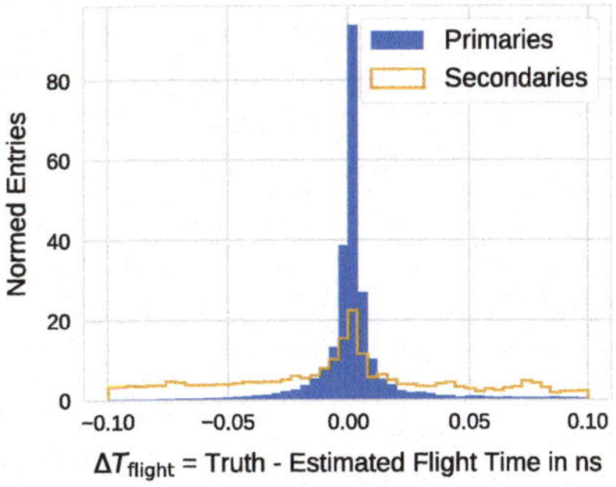

(b) Residual ΔT for particles from primary or secondary decays.

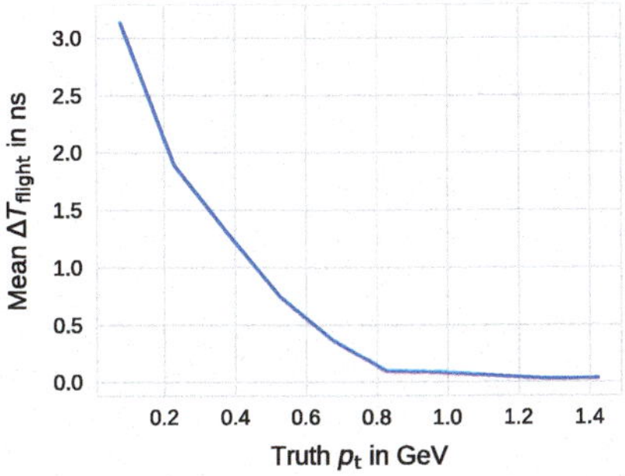

(c) Mean residual ΔT for different simulated transversal momenta of the particles.

Fig. 5.3 Deviation between the estimated flight time at each hit calculated with the procedure described in the text and the simulated flight time taken from the simulation on a generic BB event using the MC truth information during track finding

correct hypothesis. The electrons/protons being lighter/heavier than the pions arrive earlier/later at the wire leading to an overestimated/underestimated flight time for each hit. Figure 5.3b shows also the flight time estimation residuals for secondary particles, which have a strongly deteriorating resolution.

Secondary particles and wrong particle hypothesis can be discarded easily by applying the selection on the tracks described above. The reason for choosing the transversal momentum as a selection criteria is visible in Fig. 5.3c which shows the deviation of the estimated flight time from the simulated one for different particle momenta. The deviation is smaller than 2 ns for transversal momenta above approximately 200 MeV. At a transversal momentum of 350 MeV, the particles pass the CDC without curling, making the track finding and fitting simpler. The section before has already shown, that the efficiency to find such a set of hits in each event is sufficiently high even in cases of large deviations from the correct event time.

5.4 Event Time Extraction Using the Drift Length Information

If all ingredients of Eq. (5.1) except T_0 are known sufficiently well, the event time can be extracted easily. The different components together with the precision on how well they can be calculated during the reconstruction are shown in Fig. 5.4.

5.4.1 Components of the Measured Time in the CDC

The flight time is calculated as described in the previous section. Because problematic cases like curlers or tracks from secondary decays are mostly discarded by the momentum cut, the overall resolution is clearly below 1 ns. Still, there is a long but fortunately negligible tail where the truth flight time is larger than the extracted value. These wrong flight time estimations come from the problematic cases already seen in the section before, that survived the cut on the track set. However, in the vast majority of cases the deviation is clearly below 1 ns making the flight time estimation good enough for using it during the event time estimation.

The propagation time can be approximated using the z information of each hit, which is known after the track parameters are estimated using a track fit. Although the z information may not be correct in all cases, the deviation of this time component is in the order of 100 ps and is therefore negligible.

The measured time is directly given by the time counter in the detector. All signal hits have a positive measured time which produces a sharp rise in the distribution at the correct event time.

Finally, the drift time T_{drift} is the quantity which is the hardest to calculate. During the track fit, a drift time $T_{\text{drift, preliminary}}$ is calculated using the current event time hypothesis (which may be wrong), the measured time and all other time components following Eq. (5.1). The drift time $T_{\text{drift, preliminary}}$ is then turned into a drift length $d_{\text{preliminary}}$ using the measured and parameterized x–t relation which describes the drift properties of the ionized electrons in the detector gas. The track fit is performed with this drift length. Figure 5.4 shows one example of these x–t relations, which are different for each drift cell.

Another possibility, which can be applied knowledge of the event time, is to use the distance between the trajectory and the wire d_{fit}, which can be calculated using the trajectory parameters from a track fit. By inverting[3] the x–t relation, the drift time T_{drift} can be calculated independently from the event time. Both calculations are depicted in Fig. 5.5. If the current event time estimation differs from the real event time, the drift lengths $d_{\text{preliminary}}$ of each hit during the fit appear to be smaller or larger as they are supposed to be. The Kalman-based fit tries to incorporate this fact by moving the trajectory closer or farther away from the wire, which biases the

[3] As the x–t is too complicated to be analytically invertible, a numeric iterative inversion is performed.

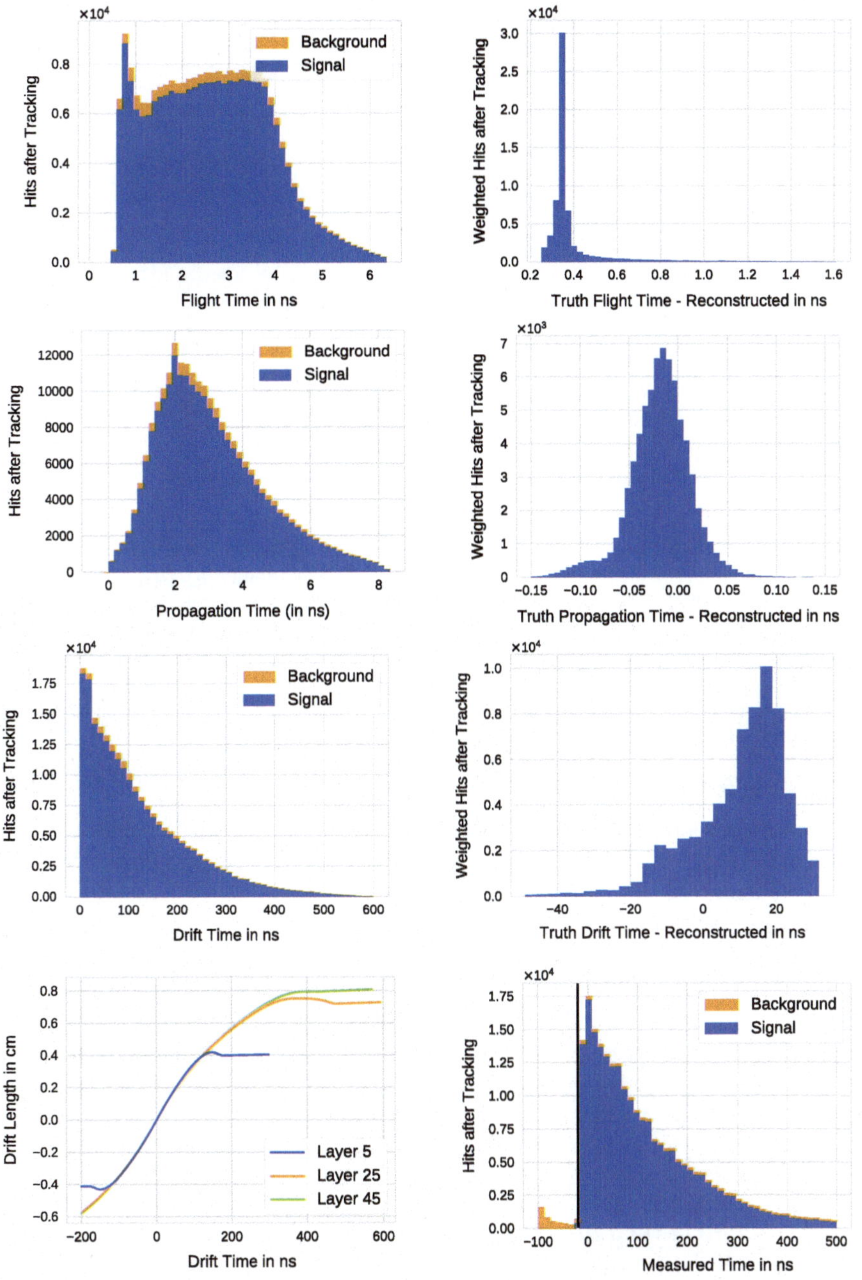

Fig. 5.4 Distribution and resolution of the different time components at each hit for simulated events with a trigger jitter of −20 ns. The resolution is weighted with the weight given by the DAF filter after fitting. Shown are only tracks with a transversal momentum above 350 GeV to cut away secondary or curling tracks. Time components. Also shown is the x–t relation which is needed for transforming the drift time into the drift length taken from first calibration measurements

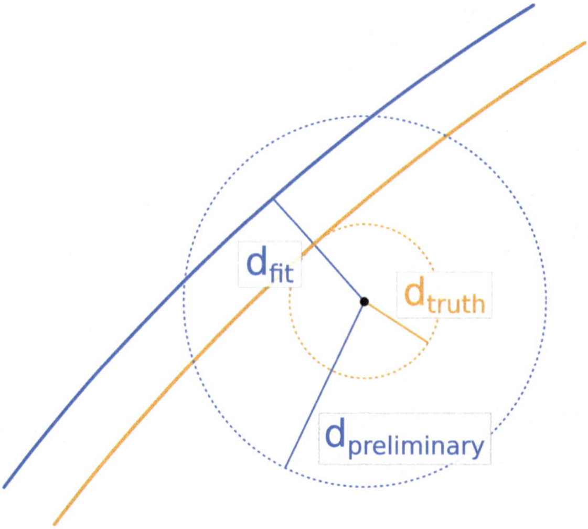

Fig. 5.5 Two different possibilities to calculate the drift length d with the truth trajectory in orange and the trajectory after the track fit in blue. The first possibility $d_{\text{preliminary}}$ uses the current event time hypothesis for calculating a drift time $T_{\text{drift, prelimiary}}$ out of the other time components T_{prop}, T_{flight}, T_0 and T_{meas} and the known x–t relation. In the shown example, the event time hypothesis is wrong and $d_{\text{preliminary}}$ overestimates d_{truth}. The second possibility d_{fit} uses the track parameters gained after a Kalman fit, which is however biased into the direction of $d_{\text{preliminary}}$ as the Kalman update tries to pull the track onto $d_{\text{preliminary}}$. However, d_{fit} can in principle be used to determine T_{drift} without knowledge of a correct T_0

calculated drift length d_{fit}. This makes the extraction of d_{fit} again dependent on the event time hypothesis, which can be seen in Fig. 5.4 showing the drift time estimation using the extracted d_{fit}. Also multiple iterations of the Kalman filter and using the deterministic annealing schema does not solve this problem.

Despite the problems in calculating the drift time using the fitted trajectory information, this method can be used to extract the event time. In each event and for each hit, the drift time T_{drift} is calculated using d_{fit} together with all other time components except T_0. The event time T_0 can then be extracted by subtracting the estimated drift time and all components from the measured time. This method was first introduced in this thesis and its performance is discussed below.

5.4.2 Performance on MC Simulations

The result of such an event time extraction can be found in Fig. 5.6. Also shown is a calculation mode using MC truth information d_{truth} for the drift length instead of the one from the track fit. The rest of the time components T_{flight}, T_{prop} and T_{meas} are the same in both modes. As it can be seen, the method itself works well if the drift

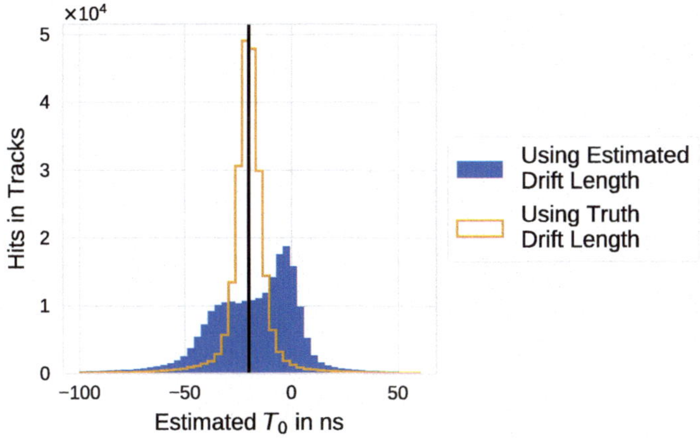

Fig. 5.6 Calculated event time using the drift length information d_{fit} extracted after a Kalman fit and using the truth information d_{truth} from the simulation. The histogram shows entries for single hits in an event set of simulated B decays for a trigger jitter of −20 ns (black line). As expected, the extracted time is biased when using the reconstructed drift length. The method itself however works, as can be seen when using the simulated drift length d_{truth} and all other parameters from the reconstruction

length information is known with a good resolution, as it is the case when using the MC truth information or the drift length d_{fit} extracted from the track fit for small values of T_0. The method fails in cases of a large deviation of the event time, as the extracted d_{fit} is biased towards $d_{\text{prelimiary}}$ (cf. Fig. 5.5).

Figure 5.7 shows the results of this time extraction method for different simulated trigger times. In each event, a weighted median of all extracted event times according to the weight given by the DAF is calculated. The procedure is repeated five times to iteratively refine the drift length information. As anticipated, the method works best for small deviations from the correct event time and achieves resolutions below the desired 2 ns. The much worse resolution with a strong bias for absolute event times larger than approximately 10 ns makes this method unusable for the full range of expected trigger time jitters. There are different possible solutions for this problem available and one of them is described in Sect. 5.7.

5.5 Event Time Extraction Using χ^2 Information

The problem of the method described above is the strong dependence of the track fit result on the (wrong) event time hypothesis. However, this dependency can also be exploited by optimizing the event time until the fitted tracks can describe the measured hits best and the two calculated drift lengths d_{fit} and $d_{\text{preliminary}}$ nearly agree. The similarity between the two values for d is hereby estimated using the χ^2 of the track fit. The method to extract the time using the χ^2 fit information of the CDC

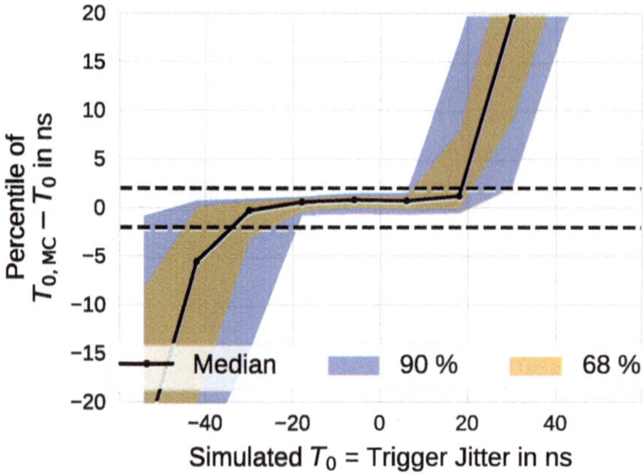

Fig. 5.7 Resolution (shown as 68 and 90% percentile) and bias of the drift time calculation utilizing the drift length information of each hit after the Kalman fit. The results were calculated on a sample of simulated BB decays including beam-induced background with different trigger jitters. The method gives precise results with moderate resolutions (90% are below 2 ns) for deviations from the event time smaller than approximately 10 ns but fails in other cases because of the biased drift length

tracks was introduced as a proof of concept earlier [7], which was an implementation of the algorithm described by [8, 9]. During this thesis, the algorithm was adapted to the tracking framework in basf2, refined and studied. The description mainly follows the discussions given in [9].

Following what was discussed in Sect. 3.3, the χ^2 value for each CDC hit in a fitted track is defined by the squared difference between $d_{\mathrm{preliminary}}$ and d_{fit} weighted by the uncertainties. The sum of χ^2 values for each track is minimal for the correct event time T_0 estimation. This can be seen in Fig. 5.8, which shows the situation for different deviations of the event time from the correct value.

5.5.1 Minimizing the χ^2 Function

For correct event times T_0 near the current event time estimation T_0', the minimization requirement of the first derivative of χ^2 can be written as

$$\frac{\mathrm{d}\chi^2}{\mathrm{d}T_0} = \left.\frac{\mathrm{d}\chi^2}{\mathrm{d}T_0}\right|_{T_0=T_0'} + (T_0 - T_0') \left.\frac{\mathrm{d}^2\chi^2}{\mathrm{d}T_0^2}\right|_{T_0=T_0'} \stackrel{!}{=} 0 \Rightarrow T_0 - T_0' = - \left.\frac{\mathrm{d}\chi^2}{\mathrm{d}T_0}\right|_{T_0=T_0'} \Bigg/ \left.\frac{\mathrm{d}^2\chi^2}{\mathrm{d}T_0^2}\right|_{T_0=T_0'} .$$

If the two derivatives are known, the correct T_0 which minimizes the χ^2 can directly be calculated. This is only possible if this time is not too far away from the current event time hypothesis. If this is not the case, an iterative procedure known as the Newton-Raphson method also discussed in [9] can be applied. In each step, the current event

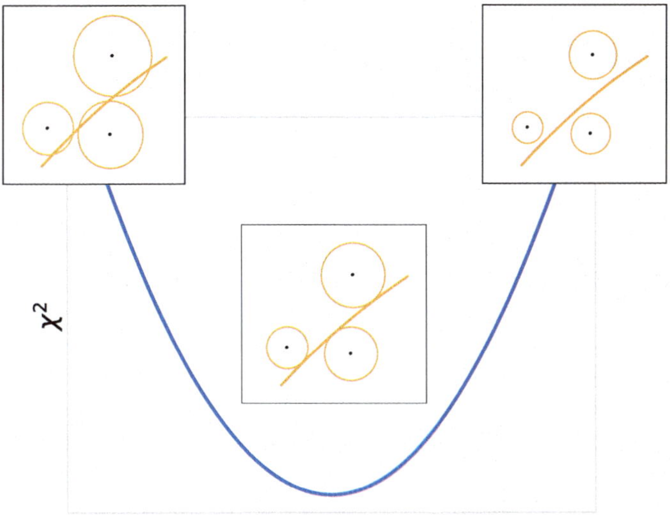

Fig. 5.8 Behavior of the χ^2 value for different deviations from the correct event time, which is zero in this case. For too large event times T_0, the extracted drift time T_{drift} is too small, because the sum of them is fixed by T_{meas} leading to an underestimated drift length. If the event time is too small, the drift length is overestimated. Both cases lead to a small distance between the track and the drift circle, which produces a larger χ^2 value for this track

time hypothesis is used for fitting the tracks in the event, which minimizes χ^2 with respect to the track parameters. Then, the next event time hypothesis is calculated using the formula above. This may not result in the exact minimum because either the linearization approximation did not hold and the event time was too far away from the correct one or the track fit was too strongly biased by the wrong event time hypothesis. In this case, the iteration needs to be repeated.

Following [8, 9], the χ^2 derivatives can be expressed as

$$\frac{\mathrm{d}\chi^2}{\mathrm{d}T_0} = 2\mathbf{A}_0^T\mathbf{V}^{-1}\mathbf{r} \qquad \frac{\mathrm{d}^2\chi^2}{\mathrm{d}T_0^2} = 2\mathbf{A}^T\mathbf{V}^{-1}\mathbf{R}\mathbf{V}^{-1}\mathbf{A}$$

where $\mathbf{V}$, $\mathbf{r}$ and $\mathbf{R}$ are defined as described in Sect. 3.3. The matrix $\mathbf{A}$ is the derivative of the residuum $\mathbf{r}$ with respect to the time T_0. In contrast to the definitions in Sect. 3.3 the covariances are required not only for a single measurement i, but for the full track. However, those matrices can efficiently be calculated using the intermediate results of the Kalman filter procedure as also described in [9]. The weights given by the DAF are taken into account during the calculation of the residuals and their derivative. An uncertainty on the extracted event time minimizing the χ^2 can also be calculated using the derivatives above with

$$\sigma_{T_0} = \left(\frac{1}{2} \frac{\mathrm{d}^2 \chi^2}{\mathrm{d}T_0^2} \right)^{-1/2} .$$

5.5.2 *Performance on MC Simulations*

Figure 5.9 shows χ^2, the first and second derivative as well as their fraction for different simulated deviations from the correct event time. The distribution of χ^2 behaves as expected: it is minimal for the correct event time and follows a nearly parabolic function near this point. In this region, only a small number of hits are downweighted by the DAF and a change in the χ^2 is basically only caused by the change in the event time hypothesis. As expected, the first derivative is approximately linear in this region.

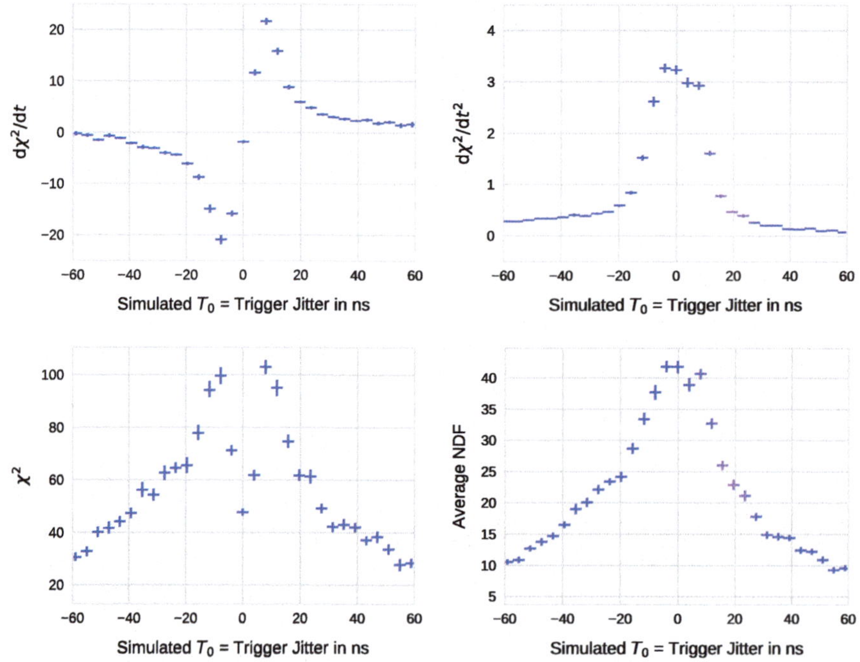

Fig. 5.9 Distributions of the $\chi^2(T)$ together with its first and second derivative and the average number of degrees of freedom in the event. For each event in the sample the average quantity is calculated using only tracks which pass the minimal p_T cut

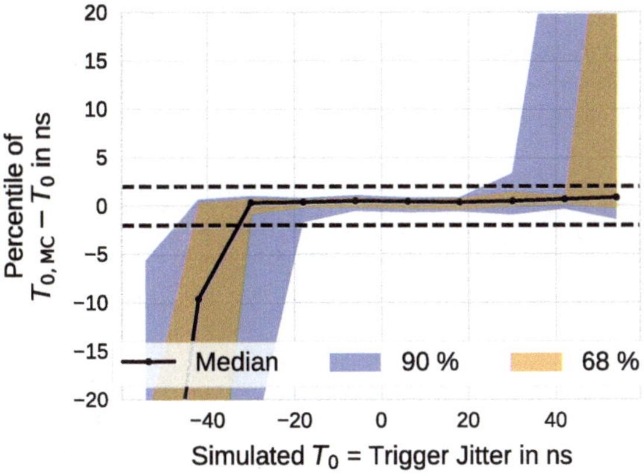

Fig. 5.10 Performance of the time estimation using the χ^2 information after a track fit analogous to Fig. 5.7. Again, the method produces very precise results for simulated event times near the correct time

It can be clearly seen that the described procedure breaks down for deviations larger than approximately 15 ns where the χ^2 distributions starts to be flat and the first and second derivatives vanish. The reason is, that the deviation from the correct event time has grown very large leading to a large residual between reconstructed position and reconstructed drift length, which is over- or underestimated because of the wrong event time. The DAF starts to discard[4] those hits because of their large residual. However, these hits are quite important for estimating the correct event time leading to an underestimated χ^2. This process can be seen in the distribution of the number of degrees of freedom also shown in Fig. 5.9.

The performance of the time extraction method using the track χ^2 is summarized in Fig. 5.10. In each event, only the tracks passing the selection criteria from above are taken into account. For those tracks, their χ^2 derivatives are summed up and the next event hypothesis is calculated using the equation above. The procedure is repeated up to five times.

As expected, the method works best for small deviations from the correct event time and leads to results with small bias and resolutions below the required 2 ns. For larger deviations however, the same behavior as seen before occurs: the resolution as well as the bias grow and make the extracted event time unusable for events with large trigger jitter.

5.6 Event Time Extraction Using Hit Information

Both of the already presented methods rely on a first rough estimate of the current event time, which must be in the region of approximately 15 ns around the correct

[4] A natural solution to this problem would be to use a Kalman-based approach without the down-weighting of hits. However this leads to track parameter estimations too bad to use the residuals reliable for time estimation.

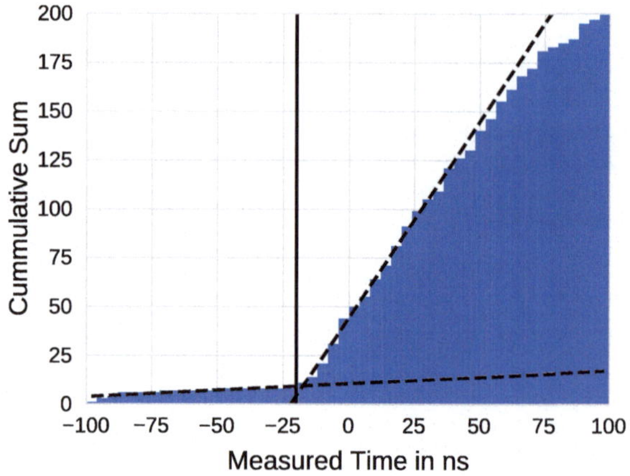

Fig. 5.11 Cumulative sum of the measured times for a single event with a simulated time of −20 ns. The kink in the cumulative distribution can be used to extract the event time using two linear functions shown in dashed black

estimate, otherwise the resulting resolution is too imprecise to be usable. Additionally, the iterative approach needed in both cases, when the estimate is not near the correct event time is time consuming. The presented approach in this section uses the bare hit information without relying on a track fit. It was developed by Thomas Hauth [4] and described here only for completeness.

As discussed, Fig. 5.4 shows the measured times for hits in events with simulated event times of −20 ns. The sharp rise of the majority of measured times in the histogram also starts at this time as expected from Eq. (5.1). The position of this rise can be determined by building the cumulative sum of the measured time distribution as shown in Fig. 5.11, while only taking into account hits which were attached to tracks during track finding to get rid of the background. The cumulated distribution has a kink at the position of the correct event time, which can be extracted using a fit of two linear polynomials to the curve.

This method operates on the bare hit information only and does not use any information from a track fit, to avoid the same problems as described above when the event time estimation is too far away from the correct time. This also means that neither the flight nor the propagation time can be correctly estimated, which leads to a worse resolution.

The result of this method can be seen in Fig. 5.12. The estimated times have a resolution in the order of 3 ns in a large region of simulated times and the resolution is mostly independent from the current time estimate. The result is biased because of additional background hits in the histogram and the wrongly estimated flight and propagation time. Additionally, the estimation may fail because the fit of the two linear functions to the cumulated histograms did not converge or the statistics in the histograms is too small to draw a reasonable conclusion. Still, the method produces reasonable results independent on how bad the initial event time estimate is. Additionally, no track fit is needed making the algorithm very fast.

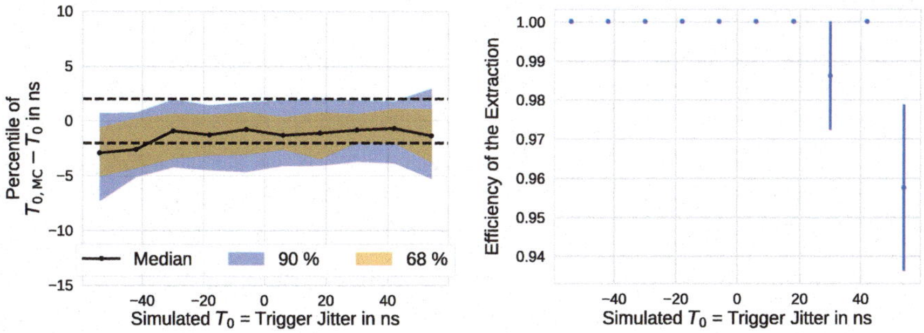

Fig. 5.12 Bias, percentiles and efficiency of the time estimation using only the hit information. The method works equally well for all trigger jitters in a large time window and produces resolutions in the 3 ns order. The bias is caused by additional background hits in the distribution. The small inefficiencies come either from failed fits of the two polynomial functions or from statistics that are too low

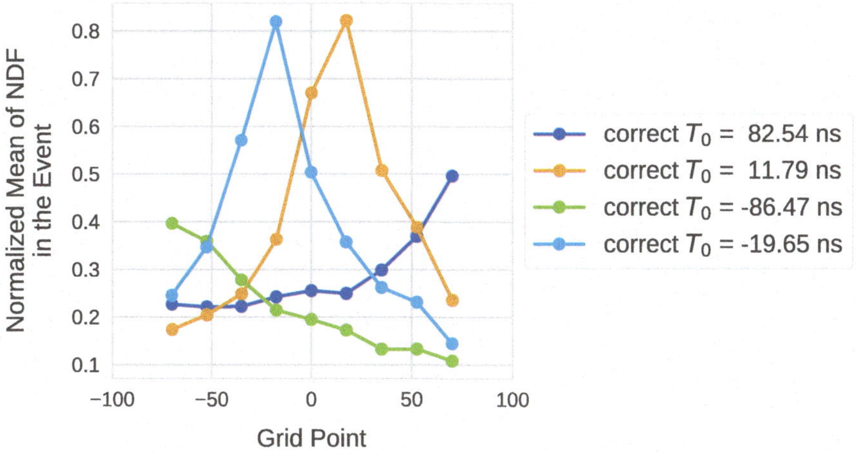

Fig. 5.13 Average NDF for four example events with different correct event times. The NDF distributions are normalized for each event to make them comparable. The marks indicate the sampling points of the grid between −70 and 70 ns. The NDF distribution is maximal near the correct event time hypothesis, even if this hypothesis does not lie in the region of the grid (dark blue and green curves)

5.7 Combination of the Methods

The presented methods can be used in combination to build a reliable and precise event extractor.

1. As a first step, the fast hit-based method is used to roughly estimate the event time T_0^s, because it is stable against large trigger jitters.
2. If the hit-based method did not succeed, a grid search is used for a good initial estimate of T_0^s. This is described in the following.

3. Either way, whether the hit-based method was successful or not, an initial time hypothesis T_0^s is now found. This initial hypothesis is refined using either the method based on the χ^2 or the with drift length d_{fit} to give the final T_0. As the χ^2 leads to better resolutions for small deviations from the correct event time, it will be used in the following. The initial estimates from the step before used for the χ^2 method is already very close to the correct event time making it possible to only use a single iteration instead of five to save processing time.

If a time-consuming grid search needs to be performed because the hit-based method failed, nine points $T^g_{i\in\{1,\ldots,9\}}$ in a large window of ± 70 ns are used as a starting point. For each grid point T_i^g, one iteration of the χ^2 method is performed. Now out of these nine extracted times, the correct initial hypothesis T_0^s needs to be extracted. Exemplary shown for a few events in Fig. 5.13, the number of degrees of freedom is maximal when the grid point lies near the correct event time. This is expected, as in the case of a correct event time, the DAF will not downweight as many hits from the tracks as if the event time hypothesis is wrong. Choosing the extracted time which maximizes the number of degrees of freedom therefore gives the best initial guess for the event time. As the χ^2 method gives precise results if the initial event time hypothesis is not farer away as 15 ns, the grid size is chosen to take this into account.

5.7.1 Performance on MC Simulations

The results of the described combined algorithm can be seen in Fig. 5.14. By construction of the mixture of hit-based and grid search, the efficiency of the combined event time extraction is 100% in all cases. The deviation between extracted and correct event time is below 1 ns for a large range of simulated trigger jitters, which is much better than anticipated. The fraction of events where the resolution is better than 2 ns is above 90% for the whole range of trigger jitters. This means that the CDC method is precise enough to determine the correct bunch crossing even without using the TOP detector in most of the cases. The largest deviation from the correct event time measured in the simulated sample was below 10 ns.

The small remaining bias in the extracted time to negative times can be explained by δ-electrons. Although they get produced on neighboring wires when the particle travels through the CDC and have therefore a finite distance to the track, they can still be picked up by the track finders. As the χ^2 algorithm tries to minimize the distance between the track and the drift circles, the event time is underestimated to increase the average size of the drift circles. To see this impact, also the extracted event times for a simulation in which the production of δ-electrons is turned off is shown in the figure. As it can be seen, the negative bias has vanished and turned into a positive bias of approximately the same size. This bias is caused by additional slower particles, which produce background hits that are accidentally picked up by the track finder and shift the average time of the track into the positive direction. This was already seen in former analyses [7]. At this stage, there is no way to erase the effect of those hits as the weight of each hit given by the DAF is already taken into account.

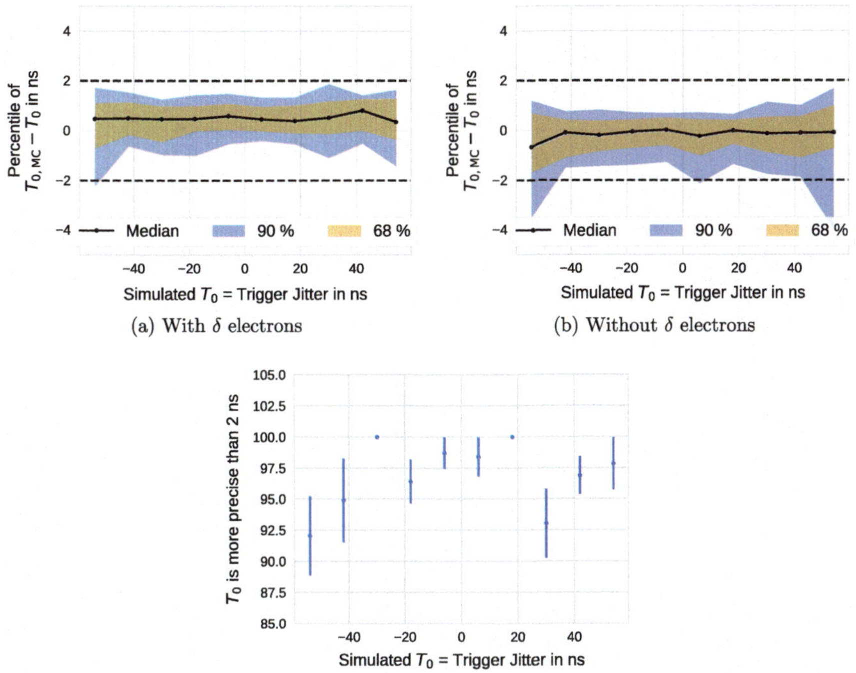

(a) With δ electrons

(b) Without δ electrons

(c) Precision

Fig. 5.14 Results of the time extraction combining the hit-based and the more precise χ^2-based algorithm with a grid search if necessary. The algorithm was tested on simulated BB decays with beam-induced background. The algorithm outperforms all other algorithms and has a resolution below the requested 2 ns for most of the cases. In only a few percent of the cases, it is not able to extract the time as precise as needed (Fig. 5.14c)

The impact of the grid-based approach in the combined algorithm can be seen in Fig. 5.15. Since the efficiency to produce an initial time estimate by the hit-based approach is high for event times in a reasonable range, the number of cases where the grid-based method is used is low. However, for larger deviations of the simulated event time from zero, the approach using a grid search becomes more important.

5.7.2 *Comparison with Other Event Time Extraction Methods at Belle II*

Besides the CDC algorithms presented in this thesis, there are additional time extraction methods using different subdetectors, e.g. the ECL or the TOP. Those algorithms are described elsewhere [3, 10] and will be used as a comparison on generated MC events as well as on first recorded collision events from Phase 2.

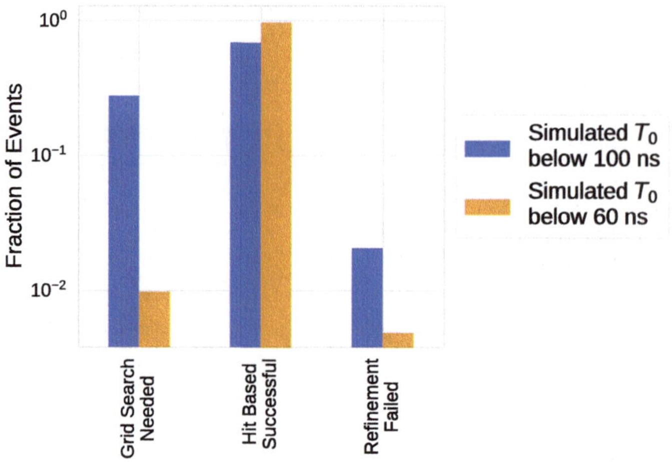

Fig. 5.15 Fraction of events where the grid-based extraction needed to run. As the search windows of the different algorithms are limited to a reasonable range of ±70 ns, the grid search is mostly needed in cases which lie outside of this range. However, also for lower simulated event times there is a low number of events where the grid-based algorithm needs to run

Figure 5.16 shows the results of the three implementations in comparison using simulated collision events including beam-induced background in the nominal Phase 3 setup. The data points indicated with CDC are calculated using the combined procedure presented in the previous section. As can be seen, the CDC algorithm outperforms the other implementations on the full shown range of trigger jitters in precision as well as efficiency. Although the ECL algorithm has also a very high efficiency in extracting an event time, its larger bias caused by clusters from beam-induced background reduces its precision. The TOP algorithm was expected to perform best. As it internally produces multiple event time candidates, it has however problems finding the best hypothesis without external input. One possible improvement in the future will therefore be to use the already very precise CDC hypothesis as input for the TOP algorithm, leading to a better selection of the final candidate.

The results presented so far were calculated using simulated collision events. The time extraction using the CDC has already been validated on the recorded cosmic ray data before the start of Phase 2 of the Belle II experiment [11, 12]. The newly recorded collision data in Phase 2 makes it possible to compare the expected behavior of the simulations with the data.

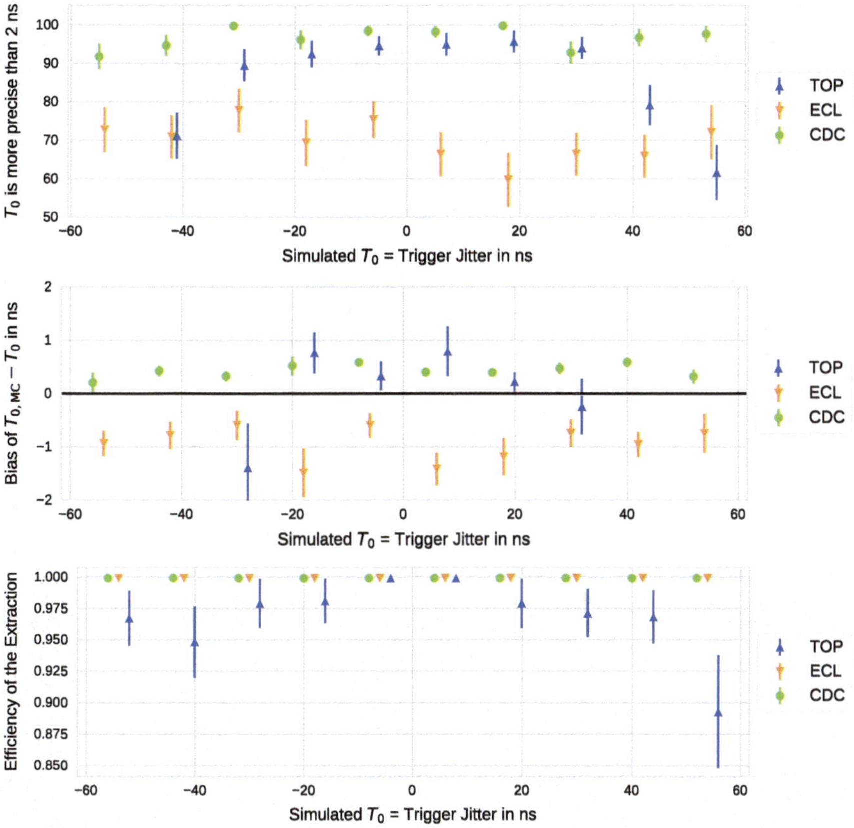

Fig. 5.16 Results (precision, bias and efficiency) of the different time extraction algorithms at Belle II in comparison using simulated BB events with beam-induced background. As expected, the TOP algorithm performs very well for small deviations from the correct event time. However, its efficiency is strongly dependent on the value of the trigger jitter. The ECL algorithm does not reach the precision of the other two implementations using CDC or TOP information. In summary, the CDC algorithm does even outperform the TOP algorithm

Figure 5.17 shows the extracted event time for events from the exemplary run 2525, where at least a single track is present. Since the correct event time is not known on data, it is impossible to calculate the resolution of the extraction, but the overall distribution looks as expected. The small bias to positive values is in line with the bias seen in the studies on MC. A comparison to the other extraction methods using ECL or TOP data is shown in Fig. 5.18, but as the calibration among the subdetectors is still missing, it is impossible to deduce the resolutions of the different algorithms.

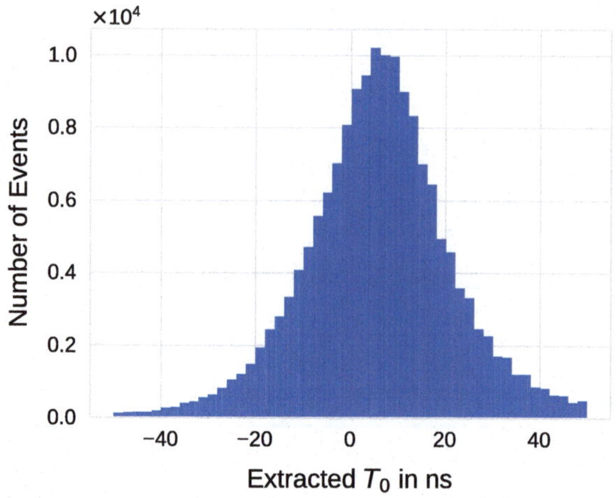

Fig. 5.17 Extracted event T_0 from all events of run 2525, experiment 3, where at least one track is present. The correct event time is not known at this stage, so it is only possible to compare the shape of the distribution with the expected behavior of the L1 trigger. The Gaussian-like shape with longer tails with a resolution in the order of 10 ns meets those expectations

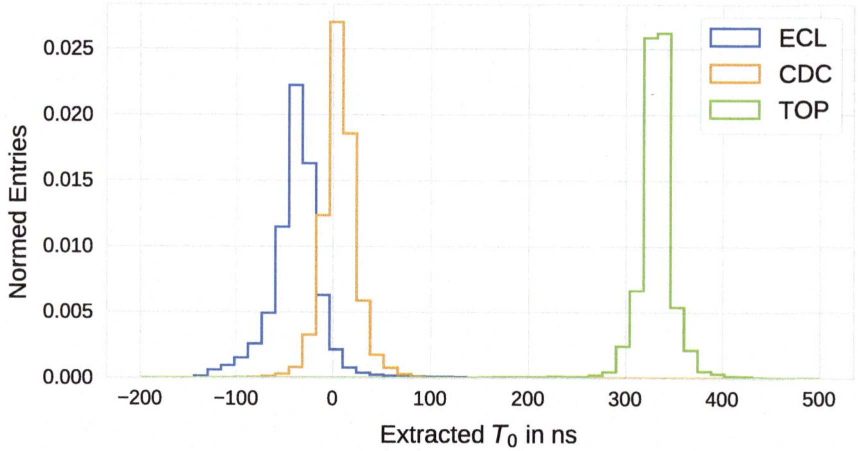

Fig. 5.18 Extracted event T_0 using different algorithms on the same events as in Fig. 5.17. It can clearly be seen, that the calibration between the different subdetectors is missing, making a determination of the resolution between the algorithms impossible

5.8 Summary

This chapter presented the first implementation of a full event time estimation based on the information from the CDC detector. Due to the large number of CDC hits, the precision in the determination is surprisingly high even (below 2 ns) surpassing the current implementation of the time extraction with the TOP. Different possible algorithms were studied and a final combination is performed, which is part of the default online and offline reconstruction. The implementation was also successfully tested preliminarily on the first recorded data of Phase 2.

References

1. Nakao M, Higuchi T, Itoh R, Suzuki SY (2010) Data acquisition system for Belle II. J Instrum 5(12):C12004. http://iopscience.iop.org/article/10.1088/1748-0221/5/12/C12004/pdf, https://doi.org/10.1088/1748-0221/5/12/C12004
2. Iwasaki Y, Cheon B, Won E, Gao X et al (2011) Level 1 trigger system for the Belle II experiment. IEEE Trans Nucl Sci 58(4):1807–1815. http://ieeexplore.ieee.org/document/5740389/, https://doi.org/10.1109/TNS.2011.2119329
3. Starič M (2017) Alignment and calibration methods for the Belle II TOP counter. Nucl Instrum Methods Phys Res Sect Accel Spectrometers Detect Assoc Equip 876:260–264. https://linkinghub.elsevier.com/retrieve/pii/S0168900217305016, https://doi.org/10.1016/j.nima.2017.04.038
4. Hauth T, Braun N (2017) T0 extraction with CDC information. In: F2F tracking meeting, Pisa. https://indico.desy.de/indico/event/19161/session/4/contribution/12/material/slides/0.pdf
5. Uno S (2018) CDC - calibration alignment performance. In: F2F tracking meeting, Tsukuba. https://indico.desy.de/indico/event/19915/contribution/1/material/slides/1.pdf
6. Eckert P Personal communication
7. Schlüter T (2016) Track-based event time determination. https://docs.belle2.org/record/338/files/BELLE2-NOTE-TE-2016-003.pdf?subformat=pdfa
8. Amoraal J, Blouw J, Blusk S, Borghi S et al (Jul 2012) Application of vertex and mass constraints in track-based alignment. arXiv:1207.4756, https://arxiv.org/pdf/1207.4756.pdf
9. Hulsbergen W (2009) The global covariance matrix of tracks fitted with a Kalman filter and an application in detector alignment. Nucl Instrum Methods Phys Res Sect Accel Spectrometers Detect Assoc Equip 600(2):471–477. arXiv:0810.2241, https://doi.org/10.1016/j.nima.2008.11.094
10. Hearty C, Ferber T (2016) ECL timing correction and resolution for simulated events in Release-00-07-00. https://docs.belle2.org/record/344/files/BELLE2-NOTE-TE-2016-006.pdf
11. Uchida M (2017) Simulation setup for CR run. In: Belle II general meeting. KEK. https://kds.kek.jp/indico/event/23987/contribution/1/material/slides/0.pdf
12. Uchida M (2017) TDC for trigger timing. In: Belle II general meeting. KEK. https://kds.kek.jp/indico/event/24065/contribution/0/material/slides/0.pdf

Chapter 6
Combinatorial Kalman Filter

The Combinatorial Kalman Filter (CKF) is a tracking concept that combines track finding and track fitting in a search-tree-based algorithm. It is used by many high energy physics experiments [1–4]—often as the main tracking algorithm. Due to its combination of track finding and fitting, it can give precise results leading to high purities as well as high efficiencies.

Due to a different detector setup at Belle II (see below), it was not clear if a CKF can be used with the same beneficial results before it was first implemented and studied in this work. Therefore, other implementations were chosen as the main track finding algorithms, e.g. see Sect. 3.2. However, employing a CKF gives valuable benefits: As the algorithm can be implemented in a way to handle hits from different sub-detectors, it can be used for cross-detector searches or to combine the results from different parts of the detector. Doing this, the implemented track finder can transfer e.g. the results from the CDC into the vertex detectors and improve the number of attached precise vertex hits to a track. Additionally, its ansatz based on the Kalman filter works very good in combination with the very precise PXD measurements. Even with high combinatorics caused by a high density of hits in a detector as it is present especially in the inner layers of the PXD, the implementation has still a small processing time.

During this work, a CKF was developed and adjusted for the Belle II experiment for a large range of use cases. The implementation covers the tasks of PXD and SVD hit attachment and merging of CDC and SVD tracks. The implementation can easily be extended to also handle different tasks, e.g. the attachment of CDC hits to SVD tracks [5] or standalone track finding.

After a motivation in Sect. 6.1, the main principles behind a CKF are shown in Sect. 6.2. The applications in the Belle II software are shown in Sect. 6.3 and the algorithm is described in detail in Sect. 6.4 with the SVD hit attachment to CDC tracks as an example. Sections 6.5, 6.6 and 6.7 are discussing the results of the implemented

N. Braun, *Combinatorial Kalman Filter and High Level Trigger Reconstruction for the Belle II Experiment*, Springer Theses,
https://doi.org/10.1007/978-3-030-24997-7_6

use cases of the CKF for Phase 3. Their performance on Phase 2 geometry is shown in Sect. 6.8. Section 6.9 summarizes the findings and gives additional applications of the CKF for track finding.

6.1 Motivation

The implemented track finding algorithms presented in Sect. 3.2 are already leading to a high finding efficiencies above 95% for a large phase-space region and purities in the same order [6, 7]. Still, there are some open problems:

- The measurements of the PXD are currently not used in the finding algorithms, as the high expected beam-induced background in those layers would lead to high combinatorics, which cannot be handled by the current implementation. Plans exist to work around this issue, however in the moment it is infeasible for the current track finders to use the PXD hits.
- The results of the SVD tracking are useful for later analysis [7]. However, it has problems when the particles do not come from the interaction point or only deposit a low number of hits in the detector (e.g. kaons or other secondary particles). Also, its performance highly depends on the background rate which is so far only estimated from simulation.
- The standalone track finding in the SVD and CDC requires an additional merging procedure. The current implementation is error prone and produces a lot of unmerged tracks, leading to miscounting of the number of tracks, the total charge or momentum etc. in later analyses. Also, the simple cuts on the distance between the extrapolated tracks used in the algorithm does not take into account the full characteristics of the independent track finding algorithms. It was already shown, that this algorithm can benefit from a more advanced implementation, e.g. using multivariate methods [8].

The CKF can solve all of these problems. The next section will introduce the basic principles behind the algorithm.

6.2 Principles

The basic procedure behind a CKF is shown in Fig. 6.1 and was first described in [9]. As the standard Kalman filter [10], the algorithm starts with a seed estimation of the track, which is created using a more basic algorithm.

Following the basic Kalman algorithm described in Sect. 3.3, the track seed is extrapolated and updated with the information of the next hit. This time however, the next hit is unknown before, as the algorithm is part of the track finding. A set of possible next hits is collected (e.g. by using geometrical properties) and the procedure of extrapolation and update is repeated for each possible candidate hit. The

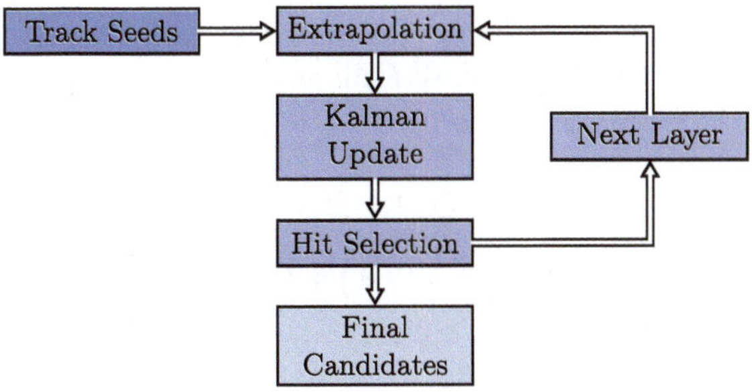

Fig. 6.1 Basic procedure behind the CKF. The three basic steps (extrapolation, kalman update, hit selection) are repeated for each possible hit on the next layer. The steps are repeated as long as there are remaining layers the track candidate can extrapolate to. The candidates that survived all filters are given to a final selection

implementation of the CKF for the specific problem defines how these candidate hits are restricted. As the time-consuming extrapolation and update procedure is repeated multiple times for each seed, the set of candidate hits is limited whenever possible by applying different filters before or after each step (summarized as hit selection in Fig. 6.1) to decrease computing time. All remaining candidates are used as seeds for the next iteration, where other hits are taken into account.[1] For each candidate, the algorithm is repeated until there are no more possible hits to attach. In the end, the CKF generates multiple candidates for each start seed. An additional selection step picks a remaining set of final candidates, where every start seed is associated to only a single candidate.

Compared to the previously described track finding algorithms already used by the Belle II framework (see Sect. 3.2), the CKF has multiple benefits. Since each candidate is inherently fitted with the full track model before a new hit is added (although there is no smoothing step), the information available at each filter step is as close as possible to the real trajectory. Other approaches, e.g. based on the Legendre algorithm, have to assume a circular trajectory. The main benefit however, is that the combinatorics is limited at each step using the full path information up to this step. Algorithms based on a cellular automaton for example need to extract useful relation information by just using 2- or 3-hit combinations. The CKF is therefore able to handle more dense environments without producing a large number of fake tracks.

This all comes with the drawback of a possibly increased processing time, as the complex extrapolation needs a lot of resources (see also Sect. 4.2). Additionally, it has problems in detectors with high hit inefficiencies as it is a local method comparable

[1]Typically, this would be hits in the next layer. Especially in the CDC but also in the regions of the VXD, where two sensor overlap, it is possible to have multiple hits per layer. Also, the next layer can either be the next layer to the inside or outside, depending on the extrapolation direction. The next section will explain in more detail, how the next hits can be found.

to the local track finding method in the CDC. These problems can be resolved by applying analogous techniques as described in Sect. 3.2 for the cellular automaton in the CDC, e.g. by allowing the algorithm to skip empty layers.

6.3 Application of the CKF at Belle II

Via the inherent track fit at each step, the CKF algorithm makes use of the spatial information of each detector measurement and is able to apply cuts based on very precise track parameters at each step. For this reason, the CKF is applied as the primary tracking algorithm at e.g. CMS [3] or ATLAS [4] experiments, starting with seeds created by combining hits in the first or last layers of the detector. These seeds only estimate the correct track parameters very imprecisely and an extrapolation to the next layer produces a larger deviation between the truth and reconstructed position. For high momentum tracks, this deviation is still small enough to select correct hits during the filter decisions while keeping the number of fakes low. With a large number of layers consisting of precise silicon sensors as e.g. in the CMS detector, it is possible to refine the imprecise seed until only correct hits are selected in the last layers and the decision which candidate to keep is simplified. If the ratio of background to signal hits is reasonable small, an iterative approach can be used: At first, seeds with a high transverse momentum and therefore lower deviations in the position estimation are used. The attached hits are removed for later iterations which decreases the combinatorics and processing time and, on the other hand, increases the purity.

At Belle II, the situation for applying a CKF is different from the previously described experiments. A seeding using the outer layers in the CDC is not feasible, as a large fraction of tracks does not reach the outermost layers or only deposits a small amount of hits there. Additionally, a single CDC hit has a limited spatial information and a large number of hits would be required, to create seeds with reasonable track parameter resolution. Using hits from the innermost layers in the PXD is however also difficult because of the high number of expected PXD clusters. As described in Sect. 2.2, the PXD measurements cannot even be read out without a prior track estimation. For seeding, there are three possibilities left, all with their own benefits and drawbacks:

- The algorithm can use seeds created with the SVD layers, e.g. by using the tracks found by the VXDTF2. Those tracks are expected to have a good spatial resolution and the VXDTF2 has a high finding efficiency on tracks originating from the IP. At the time of writing, this approach is under study by another group using the framework developed in this thesis [5].
- It is also possible to start with the innermost CDC layers, which have only a small distance to the vertex detector. The segments found by the cellular automaton algorithm in the first superlayer cover about 90% of the tracks, which gives a good starting point for the CKF. However, the first superlayer does not give any direct

z information.[2] Only by adding at least two VXD space points or a few additional CDC stereo hits leads to a first estimation of the z momentum. Together with the rather imprecise transverse momentum estimation of the seeds, this would lead to a large number of hits combinations that need to be evaluated before getting to a precise track state which can be used for extrapolation. Secondary particles that not originate from the IP play an important role for the physics performed at Belle II, so it is also not possible to restrict the z direction beforehand. Therefore, using segments as seeds is not possible.

- To get the most information from the CDC measurements, it is also possible to run the standard CDC track finding first and use the output of this algorithm as input for the CKF to attach additional SVD and PXD hits. Tracks from the CDC algorithm are expected to have a good track parameter resolution but compared to using the segments directly, the finding efficiency is reduced by about 4% caused by wrong combinations of hits and segments during the CDC finding algorithm. Additionally, wrongly added CDC hits from stereo layers can decrease the z resolution quite crucially.

Figure 6.3 shows the average number of expected space points per recorded event and also per VXD sensor for a typical BB process including beam-induced background. If not otherwise mentioned, all following studies include simulated beam-induced background as introduced in Sect. 2.1.3. At this stage of the reconstruction, the PXD hit selection is already performed. With an average number of 11 tracks per event each producing around 6 space points, the number of space points is clearly dominated by beam-induced background. This means using an iterative approach which removes the hits of previous repetitions does not help to reduce the combinatorics or improve the purity. With a good resolution in all track properties it is clearly possible to reduce the combinations to not more than 2 or 3 space points per layer. If however especially the z resolution is not high enough, it may be needed to test a large fraction of the space points in layer 5 and 6 leading to hundreds of candidates in total. Because of the small number of VXD layers, a wrong estimation of the track parameters is not refined enough to dismiss the candidate in the long run.

In conclusion, although the CKF is the standard algorithm used by many other high energy physics experiments, it is expected to behave quite differently at Belle II. Using seeds from CDC hits with worse parameter resolutions especially in the z direction impose challenges to the implementation. Additional runtime requirements of the HLT make a usage of simple heuristic cuts infeasible.

Figure 6.2 shows the default offline and online reconstruction presented in Fig. 3.2 updated with the CKF algorithms implemented in this work. Those will be discussed one after the other in the following sections. The usage of the CKF leads to a strongly increased resolution on the track parameters of the reconstructed tracks and a large

[2] In principle, the propagation time along a wire can be used for extracting the z information also for axial wires. This information however has a high uncertainty and depends heavily on a very precise knowledge of the trajectory parameter in the r-ϕ plane, which is not given for the CDC segments.

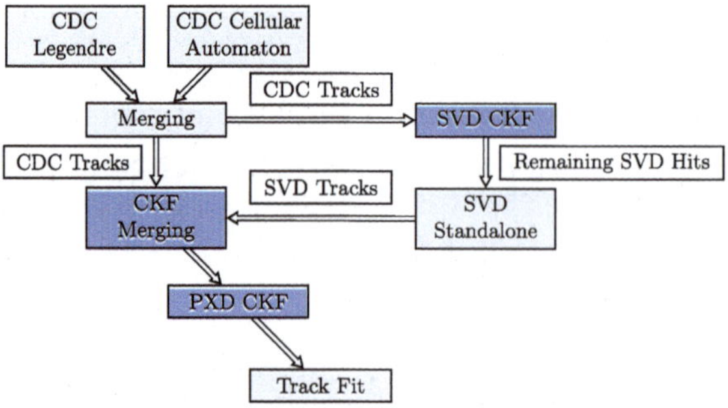

Fig. 6.2 Updated version of Fig. 3.2 including the CKF implementation developed in this thesis (highlighted in darker blue). The path shows the implementation in use for the online and offline reconstruction after this thesis

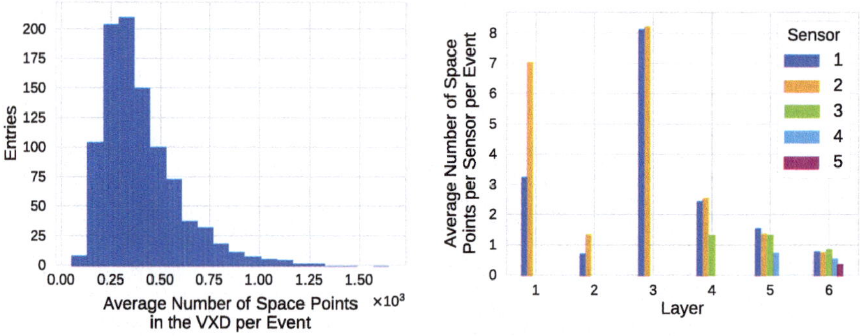

Fig. 6.3 Average number of space points (combined u- and v- strips in the SVD or clusters in the PXD) expected for a typical $B\overline{B}$ event with beam-induced background. Roughly 70 space points are expected to originate from signal tracks, the rest is beam-induced background. If the resolution of the CDC seed is not good enough, many of the space points on layer 5 and 6 need to be taken into account increasing the combinatorics

decrease in the fraction of cloned tracks. Additionally, it is currently the only track finding implementation in the Belle II reconstruction software which is able to cope with the high combinatorics in the PXD detector.

6.4 CKF from CDC to SVD—An Example

The different CKF implementations need to fulfill different requirements and it is beneficial to summarize them into a single general and broad framework. The software written for this work can handle all use cases with the same software described in the following on a single example: the CKF to attach SVD measurements to a track already found in the CDC detector.

6.4.1 Build Relations

Every basic input to the implementation is modeled as an instance of a *generic hit*. A generic hit can represent a measurement in the detector but also a track seed from another algorithm. The software is able to handle measurements from different detectors in the same way. In this example, a generic hit is either a measured SVD space point, which is a combination of two SVD clusters measured on orthogonal stripes of the same sensor or an already found and fitted CDC track. A *path* of generic hits stores the information which hits should belong to this candidate together with the current Kalman state and additional state variables, e.g. the χ^2 of the Kalman update or the weights assigned by the filters.

In each iteration of the CKF, the set of hits which should be tested next is defined by the last added hit to the path. Each hit is associated to a *box* of hits, e.g. the corresponding SVD detector sensor, and relations between different boxes are stored. On the one hand this makes it possible to handle cases with already defined layer structure (e.g. the SVD), but on the other hand also to use more complex neighboring relations which is needed in the CDC CKF. The relations among the boxes are restricted using simple geometrical constraints before the CKF starts. In the example of the SVD, relations among geometrically nearby ladders are created while allowing to skip single layers to account for hit inefficiencies. The chosen relations allow for as many correct candidates as possible while keeping the number of candidates to test computationally in reach. Depending on the applied filters, the algorithm is allowed to produce a single or multiple candidates from single or multiple hits in the box. In this example, the hits in each box are mutually exclusive and the algorithm is only allowed to add a single hit per candidate out of each sensor.

6.4.2 Tree Search

With the usage of the introduced abstract concepts of generic hits and boxes, the start of the algorithm with a seed and every following step with a detector measurement can be handled with the same algorithm. The last hit in the current path, which is either a seed or a detector measurement, unambiguously belongs to a single box. Using the hits in related boxes, additional candidates are created by adding one or more hits to the initial path. The new candidates are then filtered for obviously wrong combinations and the algorithm is repeated. In case of the SVD CKF, a new candidate is created by adding one of the hits on the next sensor to the initial path. A general overview is shown in Fig. 6.4.

The implementation is optimized for a fast reconstruction time. Additional memory allocations were avoided by reusing already created objects and the application of the strategy design pattern [11]. In total, only a small fraction of the runtime of the full algorithm is spent in the tree search framework handling the construction or garbage collection of the objects.

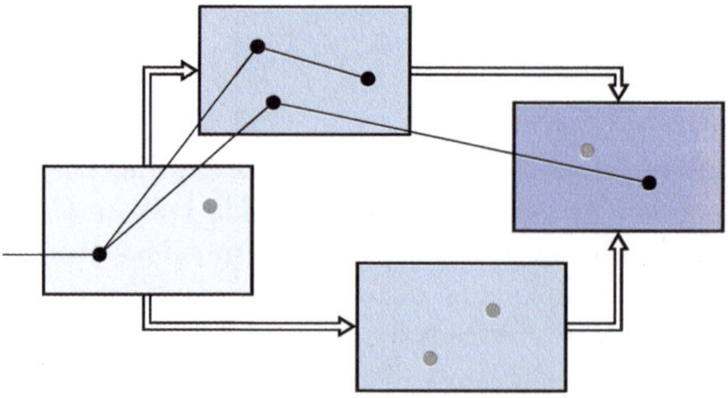

Fig. 6.4 Schematic drawing of the general setup for the tree search in the CKF implementation. Each generic hit (depicted as circle) represents a detector measurement or a seed and belongs to exactly one box (shown in blue). The boxes are related to each other (white arrows). The relations define which hits are worth considering for the next step in the path. Some exemplary paths are shown with black lines. Due to filter decisions, not all hits are used (gray circles). In each box, it depends on the implementation of the algorithm which paths get created

6.4.3 Filters

Filters are applied to the newly created path candidates to reduce the number of combinations before continuing the tree search. Hereby, it is beneficial to treat the extrapolation of the current path state and the Kalman update with the new measurement also as instances of a generic filter. They can then be mixed more easily with cut-based or multivariate filters. As described later in more detail, the SVD algorithm uses a combination of a fast filter before the extrapolation and a multivariate method before and after the Kalman update to decrease the number of candidates.

6.4.3.1 Extrapolation and Update

The extrapolation step advancing the current state to the next added hit is based on the Runge-Kutta-Nyström method as discussed in Sect. 3.3 taken from the genfit package [12]. To demonstrate the correctness of the extrapolation and the detector model, Fig. 6.5 shows the extrapolated distance between consecutive hits in Monte Carlo tracks in the PXD (first two layers) and the SVD. The distance agrees with the geometrical properties of the detector (see Fig. 2.3).

A Kalman update analogous to Sect. 3.3 is performed after the path state is extrapolated to the same plane as the measurement. The extrapolation increased the uncertainty on the track parameters. As described before, the Kalman update should reduce the residuals again as the path state is moved closer to the measurement and the uncertainty on the track state should be reduced. This expected behavior is also visible in the results of this implementation, shown in Fig. 6.6. Both these tests validate, that the extrapolation and Kalman update work as expected.

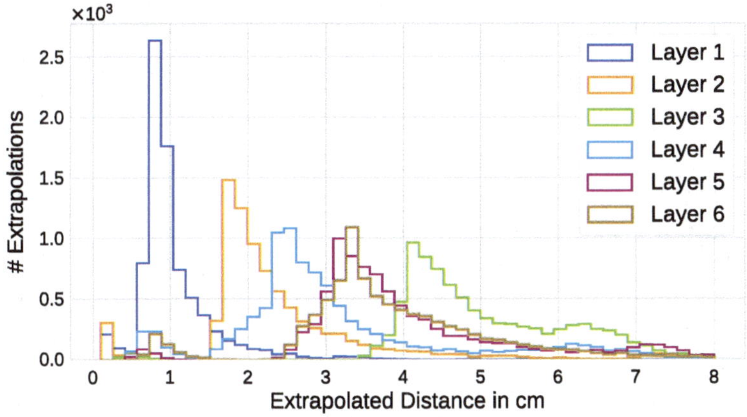

Fig. 6.5 Extrapolated distance between the hits of Monte Carlo tracks on consecutive layers. The extrapolation distances are in line with the geometrical properties of the VXD detector illustrating that the extrapolation works as expected

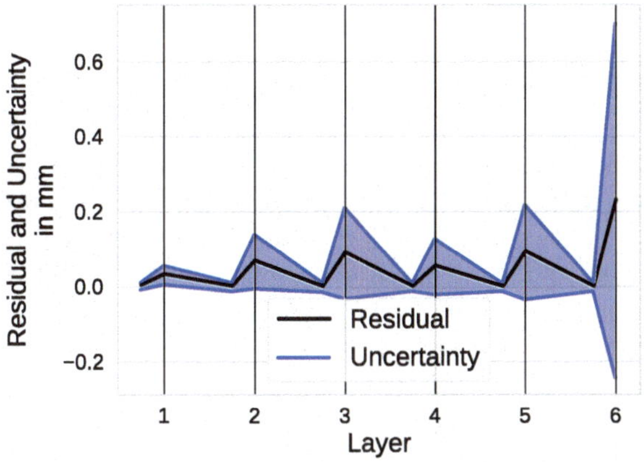

Fig. 6.6 Median residual between track and hit in each layer. For the first two layers, the residual in the two spatial hit coordinates is used, for the remaining layers only one exemplary spatial residual is shown. The uncertainty calculated by the Kalman algorithm is depicted as blue band. After the extrapolation into the next layer (black vertical lines), the residual as well as the uncertainty are large. The Kalman update reduces both quantities. Due to the refined track state, the residual shrinks approximately with each layer

6.4.4 *Final Candidate Selection*

After the tree search extracted viable path candidates for each seed using the implemented filters, a final set of candidates is produced. This is done by rating each candidate using a multivariate method taking properties of the candidate like its χ^2 or the seed properties into account. Afterwards, a set of candidates with the largest total sum of ratings with mutually exclusive track seeds is found. The SVD measurements attached to each CDC track are then merged into a final output track candidate and are fitted using a deterministic annealing filter using the full hit set of the candidate.

6.5 CKF for SVD Hit Attachment

The purpose of the first CKF algorithm shown in Fig. 6.2 is to attach SVD hits to already found CDC tracks. It replaces parts of the VXDTF2, which has the drawback of high combinatorics and requires that found tracks need to be attached to CDC tracks by an additional algorithm. The main input to this CKF implementation are found and fitted CDC tracks, which are produced by the algorithms described in Sect. 3.2. Their characteristics have been discussed in detail in other works [6] and are summarized in the next section.

6.5.1 Characteristics of CDC Tracks

Table 6.1 shows the basic track finding efficiencies of the CDC algorithm evaluated on different sets of MC particles. The efficiencies here and in the following are determined following the description given in Sect. 3.2.3. As discussed before, primary particles are generated directly during the initial collision whereas secondary particles are produced in further decays, e.g. via interactions of the primary particles with the detector material. For the physics analyses, especially the primary particles are of interest, so the tracking algorithms are optimized to have a large efficiency for primaries. The resulting tracking parameters are given near the IP at the POCA to

Table 6.1 Basic efficiencies and figures of merit of the CDC track finding algorithm calculated following the principles described in Sect. 3.2.3. Only the first outgoing arm of the tracks is taken into account. The remaining hits are either dismissed or not counted, which makes a difference in the definition of fake tracks, and is clearly visible in the numbers

In percent handling of ingoing arm	Discard	Do not count
Finding efficiency	60.9 ± 0.4	68.2 ± 0.4
Finding efficiency (primary only)	70.6 ± 0.5	78.8 ± 0.4
CDC hit efficiency	84.85 ± 0.24	82.37 ± 0.25
CDC hit purity	89.63 ± 0.08	90.02 ± 0.06
Fake rate	17.6 ± 0.4	4.18 ± 0.21
Clone rate	0.52 ± 0.08	2.97 ± 0.18

resemble the particles properties during the initial collision. Those tracking parameters are especially influenced by the found hits in the first half of the track. For this reason, in the following—if not otherwise mentioned—only the first outgoing arm of the MC track is taken into account. However, dismissing the remaining part of the MC track has an unwanted secondary effect on the finding efficiency, which is also shown in Table 6.1. In some cases, the track finding algorithm finds more hits on the second ingoing arm than on the first outgoing one. The MC track, to which this reconstructed track is compared, does not include the second ingoing arm, if one simply dismisses the hits completely. Following the description in Sect. 3.2.3, the reconstructed track will be marked as a fake track as it includes too many unmatched hits and the MC track is not counted as found. This is the reason why hits on the second, ingoing arm of the MC particles will not be simply dismissed. They are not counted when determining the hit efficiency, but also not taken into account when determining the fakes. Any hit or finding efficiency shown in the following is determined only with the first arm of the track.

As seen in Table 6.1, there is even a significant difference between the finding efficiencies with and without discarding the second half of the track. Many tracks are only counted as correct when taking into account the second half of the tracks. This means, the CDC track finding algorithm has not found the track correctly. Most likely, the algorithm has only found the second ingoing arm—even possibly confusing the charge of the particles. These tracks are a serious problem for the CKF. As the important hits near the IP are missing, these tracks have a much worsened resolution on the helix parameters.

In total, only approximately 68% of the MC tracks with five or more degrees of freedom are found by the CDC algorithms, and only approximately 79% when restricted to primary particles. Although the numbers seem to be low, they are expected when looking into the distribution of MC particles and their finding efficiency over the transverse momentum p_T in Fig. 6.7. Particles with a small transverse momentum do not deposit a large number of hits in the CDC making it very hard to find them. Especially as the first 8 layers of the CDC are axial only and particles which do not cross any stereo layer have no z information complicating a track fit tremendously. All missing particles need to be found by the standalone SVD track finding algorithm VXDTF2, which is tuned for these low-p_T cases.

Only the found and fitted tracks created by the CDC track finder can be used in the following CKF implementations as seed. To have a proper baseline, the following studies in this section will be normalized to those MC tracks, which are properly found by the CDC track finding algorithm. All studies are performed in the nominal Phase 3 setup on a sample of generic $\Upsilon(4S) \rightarrow B\overline{B}$ decays including simulated beam-induced background anticipated for the full instantaneous luminosity.

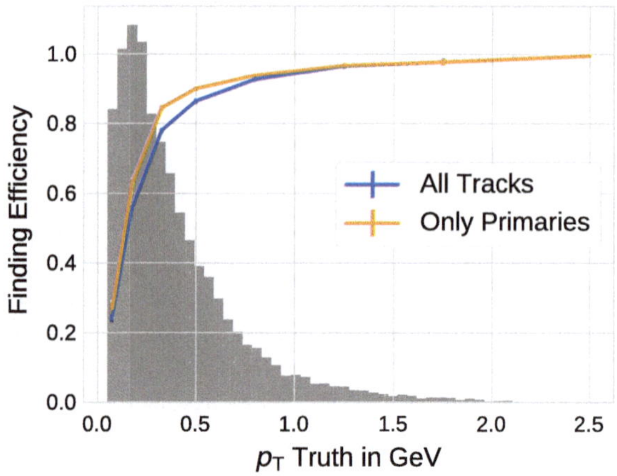

Fig. 6.7 Finding efficiency of the CDC track finding algorithm over the transverse momentum for primary or all tracks measured on a generic $\Upsilon(4S) \rightarrow BB$ sample with beam induced background. The distribution of the transverse momentum in MC is shown as a gray histogram

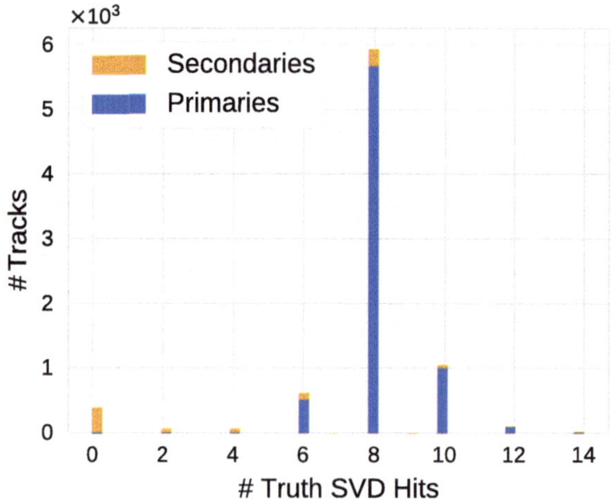

Fig. 6.8 SVD hit distribution of MC tracks, which are correctly reconstructed by the CDC track finding algorithms. As expected, most of the tracks coming from the IP deposit energy in an u- and v- strip. Only a few tracks have more hits than one per layer because of the overlapping geometry in the SVD. Secondary tracks can pass the SVD only partly

6.5.2 *Legacy Implementation*

For comparison and the definition of suitable figures of merit, the implementation without utilizing a CKF is analyzed here. This previous default algorithm will be referred to as *legacy* implementation in the following and is replaced by the work of this thesis. The algorithm uses the default CDC track finding and the VXDTF2 for creating tracks in the two sub-detectors. Two track stubs are merged based on the distance between the tracks extrapolated to the same plane between the tracking volumes [8]. Figure 6.8 shows the distribution of SVD clusters in the MC tracks. As expected, most of the tracks deposit energy in an u- and v- strip in each of the four layers, which makes it possible to use the combined space points instead of the single-dimensional clusters. Tracks not originating from the IP (secondaries) may travel only through parts of the SVD creating less hits.

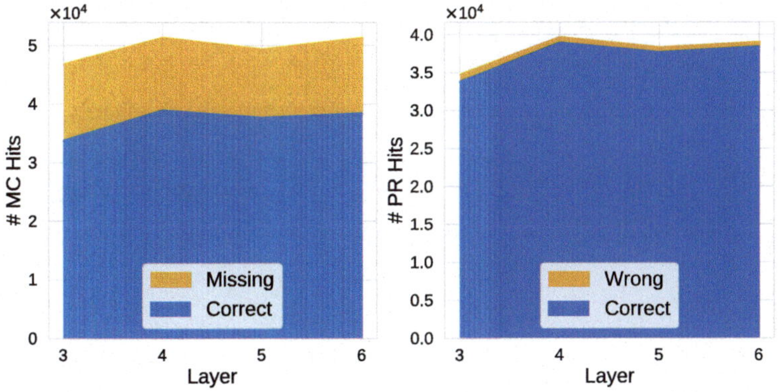

Fig. 6.9 Hit efficiency (left) and hit purity (right) for different layers in the SVD produced by the legacy implementation. Both quantities are independent on the layer number. The number of wrongly attached hits is small, but the fraction of missing hits is comparably large

Table 6.2 Basic properties of the legacy implementation calculated taking only those MC tracks into account which have at least one SVD cluster. The small number of tracks which have a SVD hit attached and the small hit efficiency already show possibilities to improve

In percent	Legacy implementation
Any correct SVD hit	81.17 ± 0.25
SVD hit efficiency	73.64 ± 0.26
With $\geq$1 found SVD Hit	90.12 ± 0.09
SVD hit purity	98.41 ± 0.07

Figure 6.9 gives the SVD hit efficiency and purity as key performance indicators of the legacy algorithm for different layers. In additional important performance numbers are also summarized in Table 6.2.

The algorithm shows a high purity independent on the layer number although the beam-induced background is higher at inner layers. As has been shown in e.g. [7], this high purity is already expected from the basic figures of merit of the VXDTF2. The small number of tracks with SVD hits attached as well as the small hit efficiency are clear indications, that either the VXDTF2 has inefficiencies in finding hits of certain tracks or the merging procedure did not combine the correct tracks. As the efficiency is independent of the layer, a systematic hit inefficiency of the VXDTF2 can be excluded.

The high number of missing SVD hits as well as the high fraction of tracks without a SVD hit at all have a large impact on the upcoming physics analyses. Approximately 20% of the CDC tracks cannot take advantage of the precise SVD measurements. Figure 6.10 shows the resolution of z_0 for tracks with and without added SVD hits. In this figure as well as in the following resolution studies the 68% and the 80% coverage of the residual between reconstructed and simulated quantity (in this case

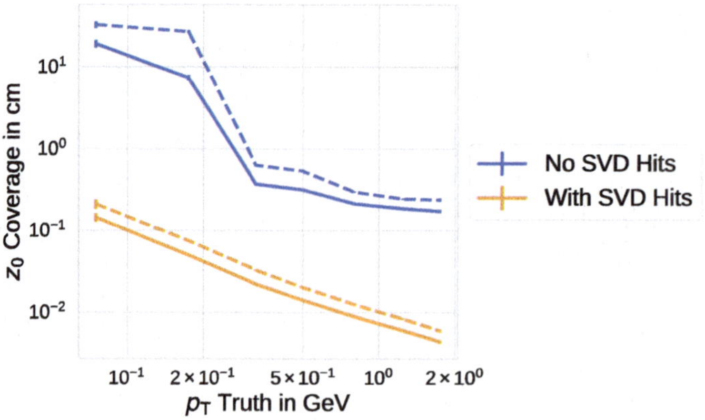

Fig. 6.10 Coverage of z_0 after applying the legacy implementation. Only tracks with at least a single CDC hit are shown separated between tracks with or without any SVD hits attached. As expected, tracks without any precise SVD information have a worsened parameter resolution. Shown here as well as in all following resolution studies are the 68% coverage (solid line) and the 80% coverage (dashed line) of the respective quantity (z_0 in this plot)

z_0) is shown. The coverage c_q is defined as the q-th percentile of the centered positive distribution

$$c_q = P_q(|X - P_{50\%}(X)|) ,$$

where X is the residual under study and P_q is the function returning the q-th percentile of the data sample. For a gaussian distribution of data points, the defined 68% coverage and the standard deviation, which is mostly quoted as the resolution, are equal by definition. In contrast to the coverage, the standard deviation is largely influenced by outliers, which can occur due to failed reconstructions. The additional shown 80% coverage indicate the shape of the resolution distribution as the full distribution cannot be shown for all cases.

Additional to the worse parameter resolution, SVD tracks which were found by the VXDTF2 but not merged to any CDC track can lead to double counting. As described in Chap. 3, a crucial cut for many physics analyses is to have no remaining tracks in the event when the full decay topology under study is reconstructed. Additional tracks can then lead to a reduced event selection efficiency.

The above mentioned problems will be solved by the newly developed SVD CKF. For this, the approximately 20% of tracks and 27% of missing SVD hits need to be assigned to the remaining CDC tracks while keeping the hit purity at a reasonable level. This implies that the CKF needs to cope with CDC seeds, that have a large deviation in their track parameters as seen in Fig. 6.10.

Table 6.3 Basic geometrical cuts used when building relations between SVD sensors and CDC tracks or between SVD sensors. The relations define which hits are used in the algorithm and are therefore very soft

Relations between CDC track and SVD sensors	
Layers taken into account	5, 6
ϕ_{max} between seed position and sensor center	$\pm 60°$
Relations between SVD sensors	
Maximal layer difference	2
Minimal layer difference (Except Overlapping Ladders)	1
Maximal sensor number difference	1
ϕ_{max} between sensor centers	$\pm 60°$

6.5.3 Performance

To understand the results of the full SVD CKF, the performance of the different building blocks which were discussed in the overview of the implementation in the previous section are investigated independently. The final result of the CKF will be presented at the end of this subsection. During these pre-studies, filters exploiting the MC truth information are used. The information is based on the MC track finder and MC matching as discussed in Sect. 3.2.3. First, the found CDC track is matched to a generated MC track. Each SVD hit which belongs to the first outgoing arm of this track is counted as correct.

Influence of the Relation Building The first step is building relations between *boxes*, which are given by the different SVD sensors. For this, basic geometrical cuts are used as defined in Table 6.3. The cuts are very loose and allow for large outliers. For faster run time, the cuts are evaluated using the current CDC seed position without performing any extrapolation to the SVD.

The results with and without this pre-cut are summarized in the first and second column of Table 6.4. All the remaining decisions in the CKF are replaced by using the MC truth information.

The hit purity is 100% in all cases by definition of the MC truth filters. Even when using no pre-filtering, the efficiency to attach an SVD hit is not 100%. Charged particles can deposit only a small amount of energy in one of the orthogonal strips below the energy threshold leading to single-clustered hit. These hits are not further used, as space points are always formed out of a combination of a u- and a v-strip. Additional, a small fraction of secondary particles, which deviate strongly from a helical track coming from the IP may deposit hits in a single layer twice on different locations. This leads to a hit inefficiency as the algorithm assumes only a single space point per layer (except for overlaps). The resulting inefficiencies are however small and can be neglected compared to the inefficiencies of the legacy implementation. As the pre-filtering is very coarse, it does not reduce the efficiency significantly.

Table 6.4 Basic properties of different CKF implementations with the same assumptions as Table 6.2. A very broad pre-filtering on the relations is implemented following the cuts described in Table 6.3. Although using MC information, the efficiency is not 100% caused by single-clustered or secondary tracks

In percent	MC truth		MC truth with pre-filter	
CDC from MC	No	Yes	No	Yes
Any correct SVD hit	99.91 ± 0.02	99.88 ± 0.02	99.91 ± 0.02	99.87 ± 0.02
SVD hit efficiency	99.54 ± 0.03	99.38 ± 0.03	99.53 ± 0.03	99.36 ± 0.03
With ≥1 found SVD hit	99.63 ± 0.02	99.50 ± 0.02	99.62 ± 0.02	99.48 ± 0.02
SVD hit purity	100.0 ± 0.00	100.0 ± 0.0	100.0 ± 0.0	100.0 ± 0.0

There is a small visible difference between seeds from MC and from the CDC track finding algorithm. The MC seeds include all particles, also tracks with very unusual parameters. The assumptions of a single space point per layer and two strips in u- and v-direction does not hold for all of them. The CDC track finding algorithm has likely not found these abnormal tracks and as the numbers are normalized to the seed tracks, the normalized efficiency increases.

Influence of the Extrapolation and the Material Effects Calculation During the application of the Kalman filter, one important step is the extrapolation from one layer to the next to add additional hits. As described, the extrapolation is handled by the Runge-Kutta-Nyström algorithm taking into account the energy loss caused by interactions with the detector material. A proper particle identification is not present at this stage of the reconstruction. The calculation of the material effects is therefore performed with a pion hypothesis for all tracks, as most of the produced particles at Belle II are pions. If this hypothesis is wrong or the track parameters deviate too strongly from the correct ones, the extrapolation fails or leads to unphysical results. In these cases the path is aborted. Additionally, the extrapolation is restricted to the inwards direction. Heavily curling tracks or tracks with a large deviation caused by multiple scattering are dismissed by this restriction. The result is shown in Table 6.5. Enabling the extrapolation leads to an unavoidable deviation in the efficiency.

For this reason, a second implementation of the extrapolation was tested without taking the material effects into account. The trivial approach leads to worse resolutions and more imprecise track parameters, but has only a small dependency on the particle hypothesis or a correct calculation of the energy loss. As will be seen later, the cuts on the parameters have to be very soft to still include a reasonable amount of outliers, so the highest precision in the tracking parameters is not beneficial. The largest benefit of this approach is its crucially decreased runtime by one to two orders of magnitude compared to the implementation with material effects.

The influence of the extrapolation is especially visible when using seeds from the CDC track finders. Wrong trajectory parameters lead to failing extrapolations and a

Table 6.5 Continuation of Table 6.4 including the required extrapolation from one layer to the next, still using MC truth filters

In percent	MC truth with pre-filter and extrapolation			
Material effects	Yes		No	
CDC from MC	No	Yes	No	Yes
Any correct SVD hit	99.24 ± 0.06	99.53 ± 0.04	99.11 ± 0.06	99.49 ± 0.04
SVD hit efficiency	96.75 ± 0.09	98.95 ± 0.05	96.65 ± 0.09	98.90 ± 0.04
With ≥ 1 found SVD hit	97.49 ± 0.09	99.41 ± 0.03	97.51 ± 0.09	99.40 ± 0.03
SVD hit purity	100.0 ± 0.0	100.0 ± 0.0	100.0 ± 0.0	100.0 ± 0.0

reduced efficiency when the extrapolation is turned on. Additionally, curling tracks found by the CDC algorithm with wrong reconstructed charge are extrapolated in the wrong direction leading to additional hit inefficiencies. The effect is not present when using MC seeds leading to a much better efficiency.

Ignoring the CDC tracks with a very bad resolution when adding SVD hits can however even be beneficial. The VXDTF2 running at a later stage might be able to reconstruct the particles properly and an additional CKF extrapolating from the SVD to the CDC can attach additional hits leading to a better resolution for those tracks. The initial CDC track can then be dismissed.

Influence of the Single Hit Filters The next step is the replacement of the MC filters with user cuts. Most of the wrong hits can be dismissed quite fast by cutting on the residual of the hit to the track or the χ^2. To save computing time especially in the time-consuming extrapolation, a pre-filtering before the extrapolation is performed. Hits on the far side of the detector seen from the CDC seed are dismissed. The final decision whether to keep or dismiss the currently added hit of the path candidate is performed using a multivariate method (MVA). A boosted decision tree [13] incorporates the correlations between the input quantities correctly, e.g. to cut softer on low-p_T tracks, which are harder to find by the CDC track finding algorithm and harder on lower layers, as here the influence of the SVD measurements on the track state is higher. Three trained MVA methods are applied before and after the extrapolation as well as after the Kalman update to include as many information as possible. They are all trained on an independent set of simulated events including beam-induced background.

The most important variable used for classification is the χ^2 value after the fit, which is supposed to be high for wrong combinations of seeds and hits. As the extrapolation uncertainty is taken into account, the worse resolution on the CDC track parameters should be handled correctly. Figure 6.11 shows the percentiles of the χ^2 distribution for correct and wrong combinations. It can clearly be seen, that seeds from the MC finder allow to cut about one order of magnitude harder compared

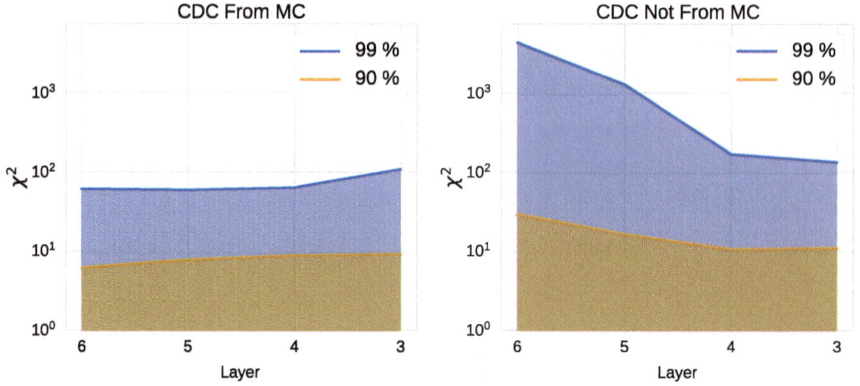

(a) correct candidates

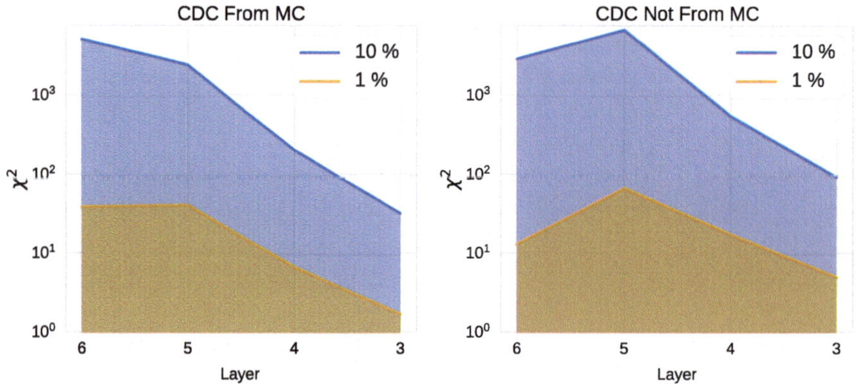

(b) wrong candidates

Fig. 6.11 Percentiles of the χ^2 distribution of correctly (top) or wrongly (bottom) added hits to the current path candidate for different layers (only showing extrapolations from one layer to the next). As the seed estimates are worse for non-MC CDC tracks, a χ^2 cut needs to be much softer accepting an order of magnitude more wrong combinations

to seeds from the CDC track finder, because of the worse resolution. Only for the last layer, the difference has vanished. This however means, to reach the same efficiency when using non-MC seeds, an order of magnitude more wrong combinations need to be accepted. Analogous conclusions can also be drawn when looking to other classifying quantities, e.g. the residuals.

The performance is shown in Table 6.6. Still, a MC filter is used to select the best single candidate for each CDC track seed in the overlap check. The tree search is performed without truth information. The MC overlap filter attaches the path candidate with the highest efficiency to the CDC seed, which obviously biases the results. To keep the efficiency at a high level, only combinations with a low classifier output are dismissed. The rather small implied efficiency drop for each single hit

Table 6.6 CKF results analogous to Table 6.4 with three MVA-based selections at each step. The final selection is still performed using a MC-truth filter. The MVA method decreases the hit efficiency compared to the MC-truth filter

In percent	MVA filter in tree search with MC truth overlap check			
Use material effects	Yes		No	
CDC from MC	Yes	No	Yes	No
Any correct SVD hits	92.14 ± 0.16	98.61 ± 0.07	91.99 ± 0.17	98.58 ± 0.07
SVD hit efficiency	90.87 ± 0.16	97.76 ± 0.07	90.57 ± 0.19	97.38 ± 0.08
With $\geq$1 found SVD hit	96.53 ± 0.11	98.97 ± 0.05	96.44 ± 0.11	98.62 ± 0.05
SVD hit purity	97.51 ± 0.09	99.49 ± 0.03	97.59 ± 0.09	99.55 ± 0.03

decision has a clear impact on the performance, as loosing hits on outer layers traverses down to inner layers increasing the overall hit inefficiency. To keep the runtime low also in cases of a large number of clusters in an event, which can happen because of showering particles in the SVD, the number of candidates to test per layer is restricted to 10. The average number of candidates per layer is only in the order of two candidates. Again, both modes with and without proper handling of material effects give similar results. However, neglecting material effects is much faster, and hence will be used in the following as default. As expected from the analysis of the input variables exemplary shown in Fig. 6.11, using MC seeds leads to a much higher efficiency in selecting the correct hit candidate at each step. It additionally shows, that the implemented MVA filters are able to find the correct hit with high precision if the seeds represent nearly the correct track parameters. As the optimization of the CDC seeds is not part of this thesis, the MC mode will be omitted from now on.

Influence of the Final Candidate Selection As a last step, the final selection algorithm is replaced by a multivariate method. The current algorithm is applied to an independent training set and the tree search is executed. For each final candidate, variables describing the full track, e.g. the number of found hits, the summed χ^2 or the difference between updated track parameters and the CDC seed is recorded. The final selection should choose the candidate with the highest hit efficiency. This implies, that the multivariate method should be able to estimate the number of correct hits instead of just classifying between correct and wrong candidates. A typical track has four to five hits making the hit efficiency a nearly discrete quantity. Instead of training a multivariate method for regression, it is therefore possible to use the well tested and fast boosted decisions trees for classification, which are also used in the rest of the track finding code. The training procedure implies the usage of several different decision trees, which are trained according to the following procedure: A decision tree t_i is trained to classify a candidate, whether its hit efficiency is higher than c_i, where i runs from 1 to 4 and c_i is [0.2, 0.4, 0.6, 0.8], respectively. By choos-

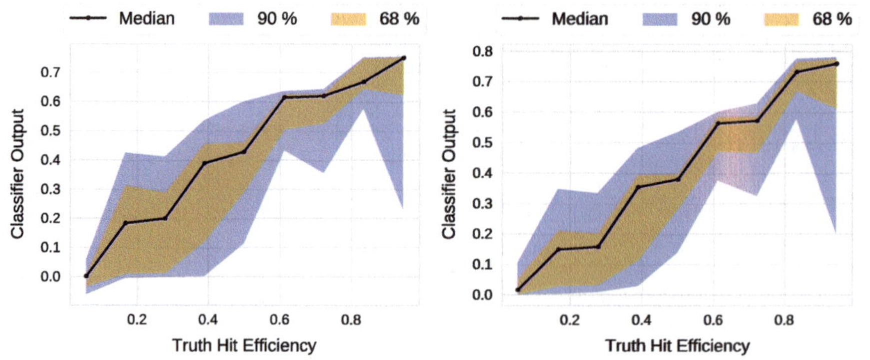

(a) TMVA Regression method (b) Combination of BDTs

Fig. 6.12 Distribution of the classifier output for the final track selection over the hit efficiency after the tree search algorithm was executed. Shown is the output of a multivariate method trained for regression (left) and a combination of five BDTs (right). Although the algorithm was not trained for a regression, the classifier output on the right resembling very closely the hit efficiency and is usable as a selection criteria

Table 6.7 CKF results without using any MC truth information in the tree search or the final candidate filtering, which makes it comparable to the legacy implementation shown in Table 6.2. All efficiencies were increased largely compared to the legacy implementation

In percent	CKF without MC truth filtering
Any correct SVD hits	89.99 ± 0.13
SVD hit efficiency	89.98 ± 0.19
With $\geq$1 found SVD hit	98.04 ± 0.07
SVD hit purity	98.31 ± 0.07

ing these c_i, it is possible to classify candidates with four- and five-hit MC tracks correctly, as the number of correct hits is always an integer number. The final decision for a candidate is given by the averaged output of all classifiers. If all classifier return a high number, the candidate is likely to have a high hit efficiency—and respectively for a low classifier output. Figure 6.12 shows the classifier output over the true hit efficiency in comparison with a method which is especially used for regression tasks from the TMVA package [14]. It can be seen, that the classifier output is very close to the target making it usable as a ordering criteria.

Final Performance A concise overview of the results of the CKF SVD without the usage of any MC truth information is shown in Table 6.7. As expected, using the MVA method as final filter has a slight impact on the hit efficiency, although the efficiency is still very high.

Comparing the performance with the legacy implementation in Table 6.2 it is clear, that the new implementation has crucial benefits. As described in Sect. 6.6, its higher hit efficiency leads to better resolutions especially in the important position parameters d_0 and z_0. As more CDC tracks have SVD hits attached, a following application of the VXD track finder will produce mostly new tracks, decreasing the number of clones. Decreasing the number of clones is important for every analysis, as it helps to clean the rest of the event after subtracting the found signal particles or corrects the expected number of tracks in an event. If one particle is found twice, it produces additional signal candidates which are hard to discriminate in the analysis.

6.5.4 Additional Runs of the CKF and VXD Track Finding

After the application of the described CKF to the CDC track set, about 10% of the CDC tracks still have no partner in the SVD. In most cases, the track parameters of the CDC seed are too imprecise to use a CKF approach in the highly occupied SVD detector. Additionally, as already seen in the beginning of this section, a fraction of tracks are not fully found in the CDC, but only parts of it. In some cases, the second ingoing arm is the only found part—sometimes even with a wrong charge assumption. The extrapolation starting with the first hit is traversing the detector in the wrong direction for these tracks.

To circumvent the previously described problem and to increase the hit efficiency for curling tracks, a second round of the CKF is applied, starting with the last hit of each CDC track as seed state. The other parts of the algorithm remain unchanged. Table 6.8 shows the performance after adding the second instance of the CKF. As expected, the algorithm is able to increase the number of tracks which have a SVD hit attached while keeping efficiency and purity at the same reasonable level.

In total, the current CKF implementation is not able to attach SVD hits to approximately 9% of the matched and properly fitted CDC tracks, which should have an SVD hit attached. When normalizing to all MC particles instead of the found and fitted CDC tracks, this leads to the results shown in Table 6.9.

Approximately 6.4% of the particles still lack an added SVD hit although they are found and fitted in the CDC. The remaining numbers are improved due to the application of the CKF or are unchanged as expected. To investigate why approximately 6% of the tracks have no added SVD hit, Fig. 6.13 shows the reconstructed charge and z_0 for these tracks in comparison with their MC truth values as well as

Table 6.8 Performance of running two instances of the CKF starting with the first and the last hit of the CDC track seed. SVD hits are attached to an additional fraction of 1% of the tracks

In Percent	CKF in both directions
Any correct SVD hits	90.88 ± 0.17
SVD hit efficiency	89.50 ± 0.19
With $\geq$1 found SVD hit	98.13 ± 0.07
SVD hit purity	98.31 ± 0.07

Table 6.9 Classification of MC particles in different categories after executing the CKF algorithm described so far. The majority of tracks have both CDC and SVD hits attached. As the remaining VXDTF2 algorithm is not applied, there are many missing tracks with no or only a small number of CDC hits, which can only be found in the SVD (category a). The CKF can only improve the situation for tracks in category (b)

MC has SVD hits	No		Yes			
Any correct CDC track found	No	Yes	No		Yes	
Any correct SVD hit attached	No	No	No	Yes	No	Yes
With CKF	12.78 ± 0.16	4.87 ± 0.10	20.07 ± 0.20	0.73 ± 0.04	6.37 ± 0.13	55.17 ± 0.25
Only CDC track finding	12.87 ± 0.17	4.78 ± 0.11	21.01 ± 0.19	–	61.34 ± 0.26	–
Category			(a)		(b)	

the distributions for p_T and $\tan(\lambda)$. Also shown are the distributions for those tracks, that have SVD hits added. For better comparison, all distributions are normalized. Many of the tracks have a lower transverse momentum and a $\tan(\lambda)$ near zero which makes them curl often in the detector. The CDC track finding algorithm is easily fooled by the distinct curls of the track and mixes up hits from different loops. This leads to worse resolutions in all tracking parameters making the tracks unusable for the extrapolation. Curlers have an additional problem which can be seen in the distribution of the charge and the z_0 resolution: Because it is not inherently clear in which sense of rotation the particle travelled, the track finder could assign the wrong charge to the tracks. This leads to large errors in the extrapolation into the SVD, as the wrong end of the track is extrapolated which can be seen in the extraordinary large deviations of the z component in the figure. As described before, the CKF algorithm is applied using the last hit as a start seed for exactly this reason—but not all of those cases can be recovered. If SVD hits are added, the charge of the track is again estimated increasing the resolution of the track parameters additionally.

In some of the aforementioned cases the standalone VXDTF2 may be able to reconstruct the tracks, as it does not rely on a correctly reconstructed CDC track. For this reason, it is beneficial to not only run the VXDTF2 afterwards to add additional low-p_T tracks but also to add additional hits to already found CDC tracks using a merging procedure. This procedure will also be based on a CKF and is described in the next section.

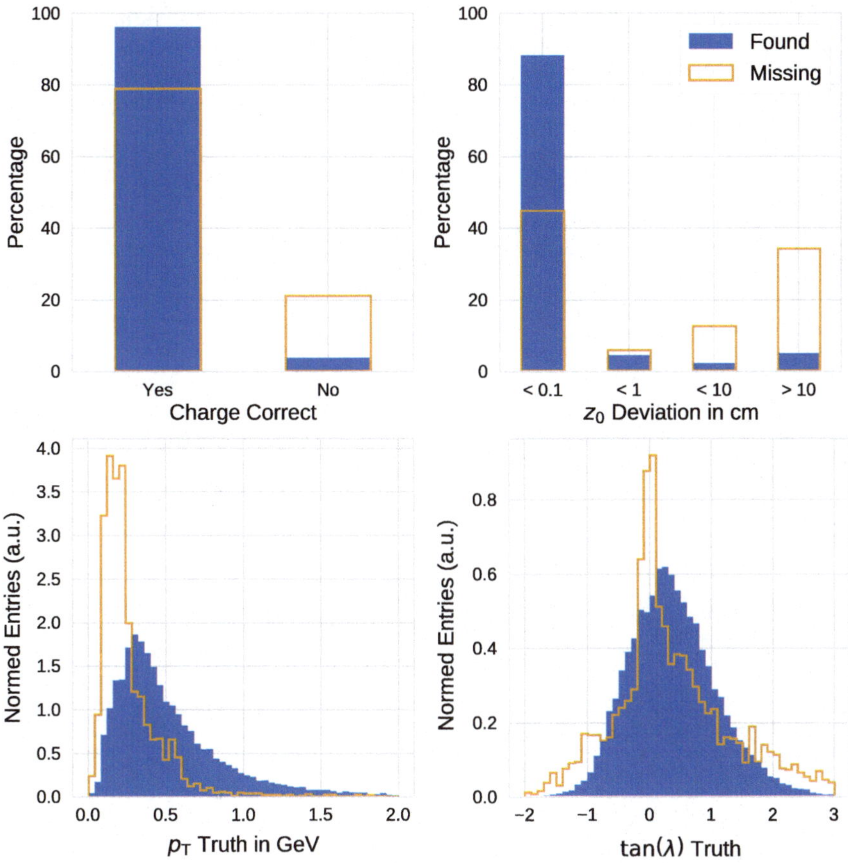

Fig. 6.13 Distributions of track parameters for tracks with no added SVD hits after applying the CKF algorithm. All shown tracks have at least a single SVD hit in MC. In comparison, the distributions for tracks with reconstructed SVD hits are shown normed to the same number of entries. Many of the tracks with missing SVD part have a wrongly reconstructed sense of rotation (charge and z_0 are wrong) and are curling tracks (lower p_T and $\tan(\lambda)$ near 0). Especially the high deviation in z_0 makes it impossible to attach SVD hits correctly, although an extrapolation starting with the outermost CDC hit is applied

6.6 Merging of CDC and SVD Tracks

As seen in the last section, the CKF algorithm so far is not able to attach hits to about 9% of the CDC tracks (or 6.6% of all MC tracks), which should have SVD hits in principle. To recover parts of the tracks and especially to find tracks which do not deposit enough energy in the CDC to be trackable, the independent VXDTF2 algorithm is run on the remaining set of SVD clusters according to the procedure explained in Sect. 3.2. The numbers shown in the following are normalized to all MC tracks and not only to those found by the CDC track finder. Table 6.10 shows impor-

Table 6.10 Basic figures of merit for the legacy implementation and the CKF with and without added VXDTF2. The numbers are extracted as described in Sect. 3.2.3. As the VXDTF2 operates on a pruned hit set after the CKF, it is able to find an additional amount of SVD-only tracks compared to the legacy implementation. The clone rate is strongly increased compared to the number without additional VXDTF2, when no merging procedure is applied. The SVD hit efficiency is calculated on all MC tracks with at least a single SVD hit, which means the CKF-only implementation has a disadvantage as all low-p_T SVD tracks are counted with zero hit efficiency

In percent	Legacy	CKF	CKF and VXDTF2
Finding efficiency	78.69 ± 0.19	63.37 ± 0.23	79.32 ± 0.21
Only on primaries	94.83 ± 0.12	76.08 ± 0.25	95.53 ± 0.12
Clone rate	10.84 ± 0.15	2.67 ± 0.10	7.63 ± 0.14
Fake rate	6.20 ± 0.13	3.88 ± 0.11	6.12 ± 0.12
SVD hit efficiency	72.06 ± 0.22	66.57 ± 0.24	83.98 ± 0.17
With ≥ 1 found SVD hit	77.62 ± 0.18	88.09 ± 0.19	89.65 ± 0.14
SVD hit purity	98.81 ± 0.05	99.00 ± 0.05	98.99 ± 0.04

tant tracking figures of merit for the legacy implementation and the described CKF algorithm so far with and without additional VXDTF2. As expected, the additional execution of the VXDTF2 recovers the finding efficiency to at least the same level as in the legacy implementation.

Compared to the legacy implementation, the finding efficiency is increased by approximately 1%. Due to removing hits of already found tracks in the SVD, the VXDTF2 algorithm is able to find tracks more easily on this pruned hit set. In addition, the number of correctly found tracks is increased since short CDC tracks with additional assigned SVD hits have now enough hits to be counted as found. The fake rate of the implementation with the CKF and the additional VXDTF2 is on the same level as in the legacy implementation, which demonstrates that the CKF did not create invalid tracks. The SVD hit efficiency is increased, as discussed in the section before. The benefits of the added VXDTF2 come with an increase in the clone rate compared to the mode without, as there are still some tracks which are both found by the CDC and the SVD track finders. An increase of the clone rate by 5% when executing the VXDTF2 is in line with the approximately 6% of MC tracks, which were found by the CDC algorithm but not used in the CKF. These additional tracks need to be merged to their CDC partner to remain a small clone rate, which is crucial for later physics analyses.

For this, the already described CKF framework is reused for a merging procedure, which is the second CKF part shown in Fig. 6.2. Each CDC track, which has no SVD hits attached already, is paired with each found SVD track by the VXDTF2 and the combination is evaluated using a BDT. The multivariate method uses similar variables as the final selection for the SVD CKF described in Sect. 6.5. It is only trained as a classification instead of a regression as there is a binary information if a CDC and a SVD track relate to the same MC particle. Finally, for each CDC track the single best SVD candidate is extracted and the hits added to it. In the remaining section, the performance of the new track finding chain for the SVD consisting of the CKF for attaching SVD hits to CDC tracks, the VXDTF2 and the CKF for merging is discussed.

6.6.1 Overall Performance of the SVD CKF Tracking Chain

Table 6.11 shows a concise summary of the achieved performance and is now the final result of the SVD CKF chain with additional VXDTF2 run. The table also shows an implementation without using any CKF, where the tracks found by the CDC and SVD track finders are merged together if they relate to the same MC particle. It therefore uses MC truth information which is not available on data. Still, it resembles the best achievable result when not using a CKF and the results from the CDC track finder and VXDTF2 as input. As the merging CKF finalizes the SVD part of the implementation, the important figures of merit will be discussed in greater detail below.

Finding Efficiency As already seen before, there is an improvement of the track finding efficiency, as the VXDTF2 algorithm is operating on a pruned hit set. In principle it would also be possible to start with the VXDTF2, merge the found tracks with the CDC tracks and run the CKF afterwards. However, as will be discussed in the next subsection, the CKF is able to produce tracks of higher quality. Apart from that, the CKF has no benefit in running on a pruned hit set—in contrast to the VXDTF2 which profits from less combinatorics and a higher purity.

Table 6.11 Figures of merit for the final CKF SVD implementation in comparison with merging the CDC and SVD tracks using MC truth information with any CKF. Compared to the legacy implementation in Table 6.10, all important quantities improved or are at the same high level. The implementation using MC matching is able to merge more correct tracks together decreasing the clone rate. As the VXDTF2 does however need to run on the full hit set, is is not able to find the same amount of tracks as when running after the CKF. Also, the CKF implementation produces SVD tracks with a better quality as shown in the next subsection

In percent	CKF and VXDTF2 with CKF merging	CDC TF and VXDTF2 with MC merging
Finding efficiency	79.10 ± 0.14	78.59 ± 0.21
Only on primaries	95.32 ± 0.09	94.63 ± 0.12
Clone rate	5.87 ± 0.08	4.21 ± 0.11
Fake rate	6.23 ± 0.09	7.02 ± 0.14
SVD hit efficiency	85.37 ± 0.11	78.17 ± 0.18
With ≥ 1 found SVD hit	91.31 ± 0.09	84.81 ± 0.14
SVD hit purity	98.97 ± 0.02	99.33 ± 0.02

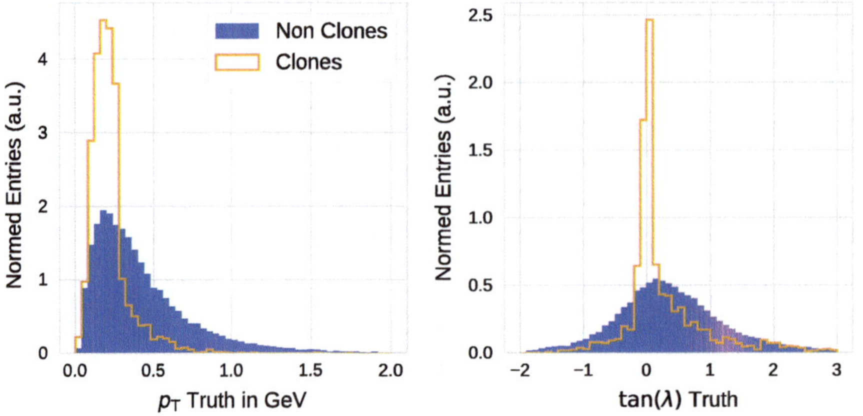

Fig. 6.14 Distribution of clone tracks after applying the described SVD CKF with merging. The distributions are very similar to the ones shown in Fig. 6.13, which is expected. Most of the clones and tracks with missing SVD hits are curlers, which are harder to find correctly due to their different loops in the detector

Clone Rate Compared to just adding the VXDTF2 tracks without any merging procedure after the CKF the clone rate decreased by about 2%, which shows that the merging procedure works in principle. The difference to the best-achievable 4.21% with MC merging however demonstrates, that the merging still does not work perfectly. The parameter distribution of clone tracks is shown in Fig. 6.14. It is similar to the distributions of tracks, which have no SVD hit attached after the CKF application in Fig. 6.13, although the shown clone tracks are mostly SVD only tracks.[3] The problems are the same: curling tracks can be found multiple times even mixing hits from different loops, which makes the extrapolated position very imprecise. Compared to the legacy implementation however, the clone rate is still nearly halved. As described before, this has a direct impact on mostly all later physics analyses.

Fake Rate The fake rate shown in Table 6.11 is basically unchanged between the legacy implementation and the CKF version. Figure 6.15 shows the 6.23% of fakes classified whether they have CDC and/or SVD hits.

It can be seen, that the vast majority of fake tracks arises during the standalone track finding in the CDC or the SVD and not during the merging process. A large fraction of fakes which both have CDC and SVD hits are caused by decays in flights. Most of the decays in flight happen at high momenta of the particles. The charged decay product has therefore approximately the same trajectory parameters as its parent. When comparing the result of the algorithm to the MC information, in which

[3]The reason for this is given by the MC matching procedure: If there are two competing reconstructed tracks for the same MC particle, the one with the higher efficiency will be called matched, the remaining candidate will be the clone. As most of the particles deposit more hits in the CDC, the found part in the SVD detector is likely to be classified as clone.

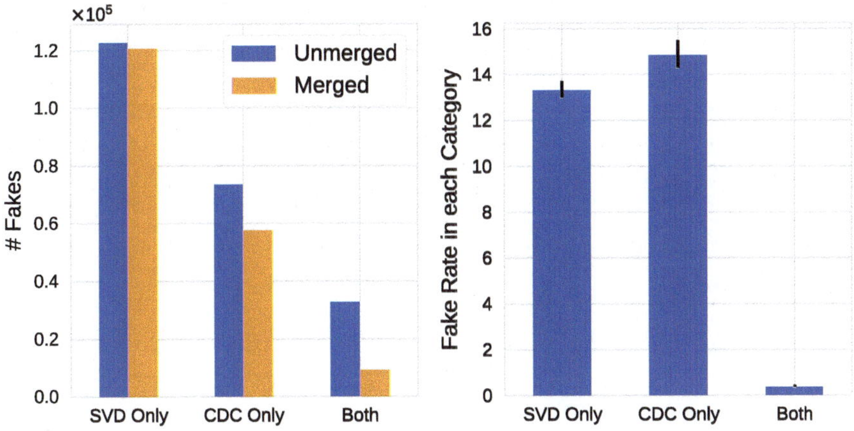

Fig. 6.15 Distribution of fake tracks after applying the described SVD CKF with merging. Most of the fakes originate from the two separate standalone track finders and are not merged together by any algorithm, which demonstrates that the CKF is able to distinguish between correct and wrong tracks. Two modi of MC matching are shown in the left figure: either the products of decays in flight are counted as new distinct tracks (*unmerged*) or are merged to their parent particle (*merged*). In the latter mode, the number of fakes drops by another 0.3% points of all found tracks. This is because many decays in flight happen at higher momenta which leaves the trajectory of the child particle unchanged with respect to the parent

decays in flight are counted as a single track instead of two separate ones, the fake rate is decreased by another 1.1%. A dedicated analysis of the trajectory shape is needed to find the decays in flight in the reconstructed tracks. It depends strongly on the later physics analysis, if handling both the parent and the decay product as a single track leads to a degradations in the particle identification or the resolution. All shown studies in this thesis are performed with keeping both particles apart in the MC truth information.

The plot shows also the fake rate in the different categories. If a track has both SVD and CDC hits, the probability of being a fake is lower than 1%, which makes this set of tracks extraordinary pure.

Hit Efficiency The largest impact of the new CKF method compared to the legacy implementation is given in the SVD hit efficiency as expected from the last section. It has improved by about 13%, which will have a clear impact on the resolution shown below. Figure 6.16 shows the SVD hit efficiency over the p_T of the particles as well as for different number of SVD hits.

As the CKF is not limited to tracks with three or more space points (six or more hits), it improves the efficiency especially for lower number of SVD hits. Although a single space point does not give much information, the CKF algorithm is able to find the hit in about 50% of the cases, therefore improving the situation especially for secondary particles (see also in the next subsection). As SVD tracks which pass the same layer twice via overlapping sensors give additional information to the CKF, the

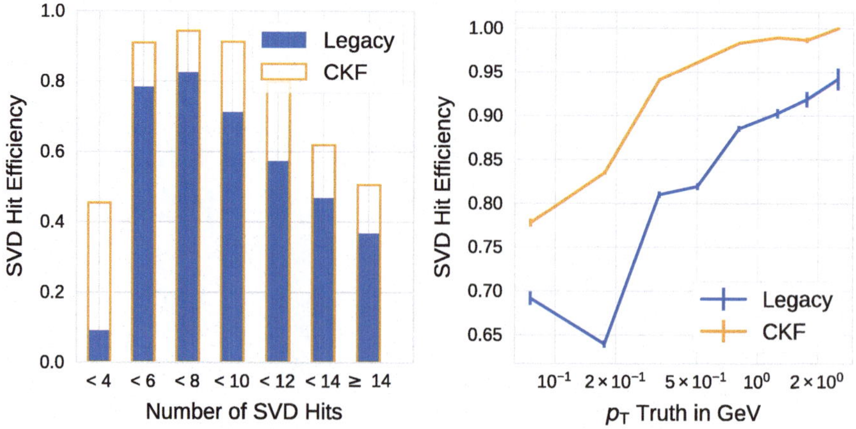

Fig. 6.16 SVD hit efficiency for different number of SVD hits and over the true transverse momentum. The CKF implementation is able to improve the overall hit efficiency. Especially for the important cases of a low number of SVD hits as well as a low transverse momenta a large enhancement is achieved, as the CKF does not need a minimum number of SVD clusters or and has less problems with low-momentum fakes. As it searches for hits in both directions of the tracks, it is able to find additional hits of curling particles between 100 and 300 MeV

hit efficiency also improves for tracks with more than 8 SVD clusters. The SVD hit efficiency of tracks in the lowest bin of the transverse momenta are dominated by the results of VXDTF2. The legacy merging and the VXDTF2 have however problems in finding or merging together tracks with medium-sized p_T, which can be seen in the dip of the hit efficiency. Here, the number of possible combinations increases drastically and with it the possible fake rate, so harder cuts need to be applied. The CKF does not show these issues.

All shown figures of merit improved compared to the legacy implementation when using a CKF—some even beyond the level that would be possible if merging the track candidates from the VXDTF2 to the CDC tracks with MC information.

Resolutions Apart from the efficiencies, the most viable output of the track reconstruction are the resolutions of the track parameters, which are shown in Fig. 6.17 in comparison with the legacy implementation. For better visibility, the improvements are also shown summarized in Fig. 6.18.

Due to the higher number of tracks with SVD hits attached, the CKF algorithm is able to improve the resolution especially for the important z_0 and d_0 helix parameters. The parameters do not only play a major role during vertexing, which needs to be performed in the majority of physics analyses, but are also a viable input on its own to analyses like measurements of the time-dependent CP violation. The improvement is especially visible for lower and medium transverse momenta, where the resolution is mostly influenced by the combination of CDC and SVD tracks. Also the momentum parameters as $\tan(\lambda)$ or p_T are improved, although the influence of the SVD hits is

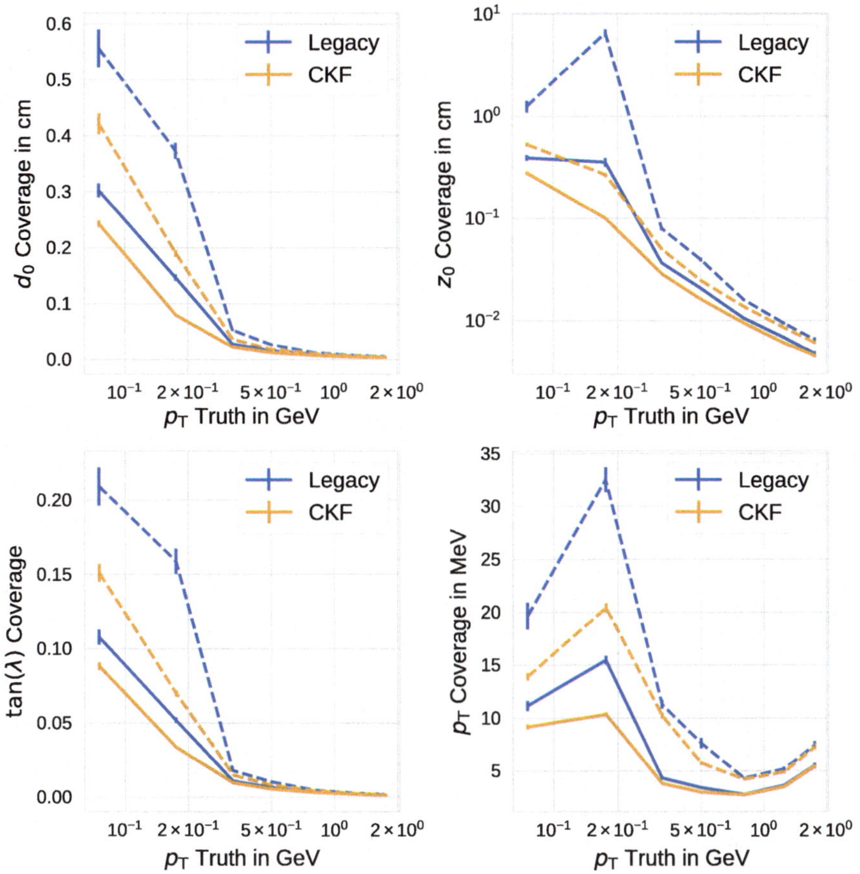

Fig. 6.17 Resolution of the track parameters d_0, z_0, tan λ and p_T after extrapolating the tracks to the point of closest approach to the interaction point as described in Sect. 3.3.1. Shown is the 68% coverage in thick and the 80% coverage analogous to Fig. 6.10. The z_0 component is plotted logarithmic for better visibility. The CKF algorithm is able to improve the estimations of all track parameters for a large range of transverse momenta, especially for the very important z_0, which is needed for analyses of CP violations. The improvement also leads to an enhancement of the very important vertexing used in every physics analysis

smaller here. As seen before, CDC tracks tend to have a worse resolution in the z direction, so additional SVD measurements can improve the tracks.

Runtime The overall processing time is an important requirement for a new track finder. Figure 6.19 shows the average processing time for the legacy and the CKF implementation. Although the CKF includes several extrapolations and applications of multivariate methods, the processing time per event did only increase by approximately 20 ms. The runtime was measured on a machine with Intel(R) Xeon(R) E5-2650 CPUs in single-processing mode. Together with the rest of the track find-

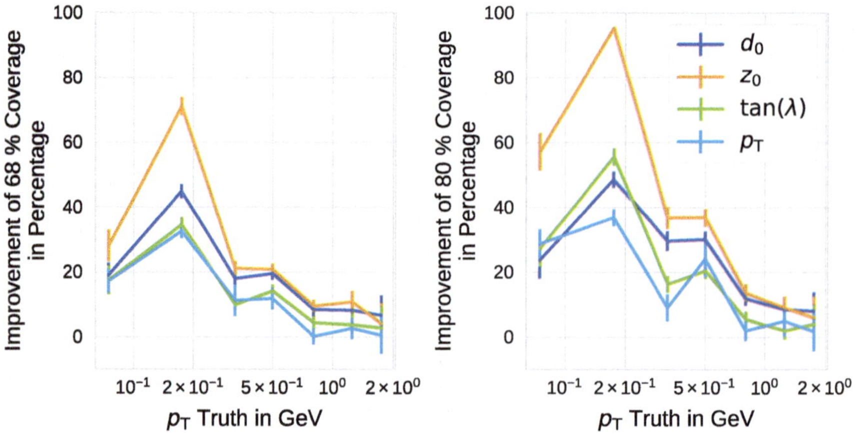

Fig. 6.18 Quotient between the resolutions after the legacy and the CKF implementation show in Fig. 6.17 for visual guidance. The track parameters are improved in the whole range of transverse momenta, but are most crucial for lower p_T

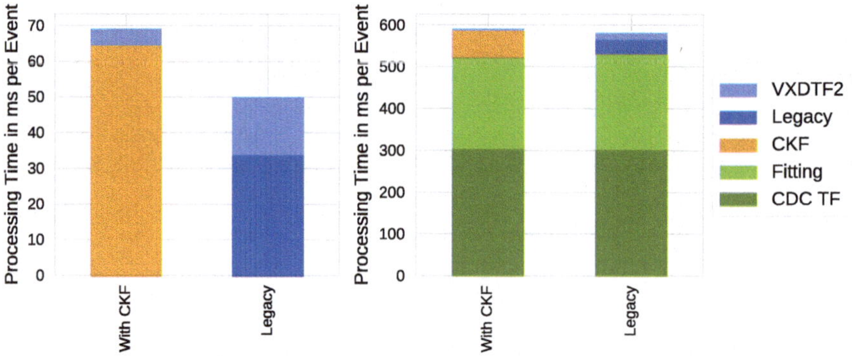

Fig. 6.19 Processing time of the legacy implementation and the new implementation using the combination of CKF and VXDTF2 in comparison. The processing time was measured on a machine with Intel(R) Xeon(R) E5-2650 CPUs in single-processing mode, which are comparable to the processors utilized in the HLT machines. The overall increase in processing time for the tracking reconstruction is only 1.7%, although the CKF modules itself need more time as the legacy merger, as the runtime of the fitting as well as the VXDTF2 was decreased

ing and fitting steps, the overall increase in runtime is only 10 ms, as the tracks are already fitted by the CKF and the final DAF does not need as many iterations as before. The fitting step does not include the fit with multiple particle hypotheses, which will be performed at the end of the offline reconstruction, but is not part of the software trigger on the HLT. Additionally, as the VXDTF2 is operating on a smaller hit sample, its runtime also decreased drastically. Even for the tight requirements of the HLT an increase of 1.7% is manageable.

6.6.2 Comparison with VXDTF2

As shown in the last sections, the new implementation using a CKF to find SVD hits and to merge VXDTF2 tracks with CDC seeds has clear benefits compared to the legacy implementation. The reasons for this are however not only given by the superior merging procedure, but also by the CKF SVD hit finding method itself. To demonstrate this, the described CKF method consisting of the SVD hit attachment and the merging of VXDTF2 and CDC tracks is compared to the VXDTF2 alone. The VXDTF2 tracks are merged to CDC tracks if they are related to the same MC track. Only properly matched and fitted CDC tracks are taken into account. The results are shown in the last column of Table 6.11.

The CKF method is able to attach correct SVD hits to $91.9 \pm 0.2\%$ of the CDC tracks whereas VXDTF2 has only found an SVD part in $84.6 \pm 0.1\%$ of the cases, although it uses MC information for merging. When restricting to a set of tracks where both methods have found hits, the CKF method still attaches $8.4 \pm 0.1\%$ more hits to the tracks. This means, the CKF method is not only able to attach hits with a better efficiency, but also attach SVD hits to CDC tracks, which would have no SVD information otherwise. This can also be seen in Fig. 6.20, which shows the distribution of hit efficiencies and the number of correctly attached hits. The CKF method finds all the correct SVD hits in $87.3 \pm 0.2\%$ of the cases, the VXDTF2 method only in $56.4 \pm 0.3\%$. In many cases, the VXDTF2 tracks lack one or two space points (2 or 4 hits), which were assigned correctly by the CKF method. As seen in Fig. 6.21, the VXDTF2 especially looses hits in the very important innermost layers of the SVD. Those hits do not only play a major role in defining the d_0 and z_0 resolution, but also define how well PXD hits can be added to the tracks. Although

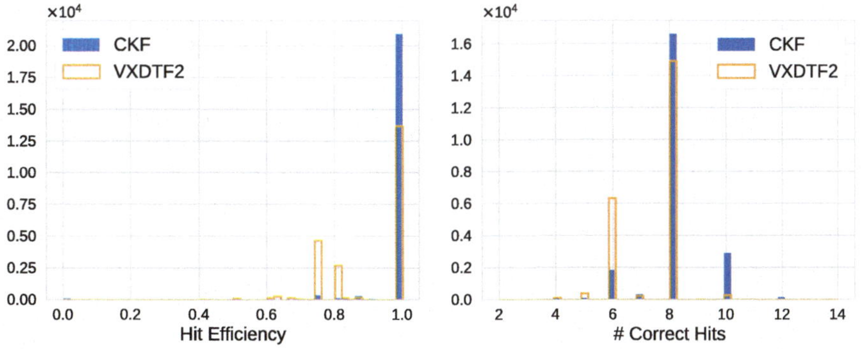

Fig. 6.20 Distribution of the hit efficiency and the number of correctly assigned hits per track for the described CKF method in comparison with the standalone VXDTF2 with MC matching applied. Although the comparison method uses MC information, the CKF method is still superior in the hit assignment. As most of the particles deposit energy on both the u- and v-strip, the distinct peaks at even number occur

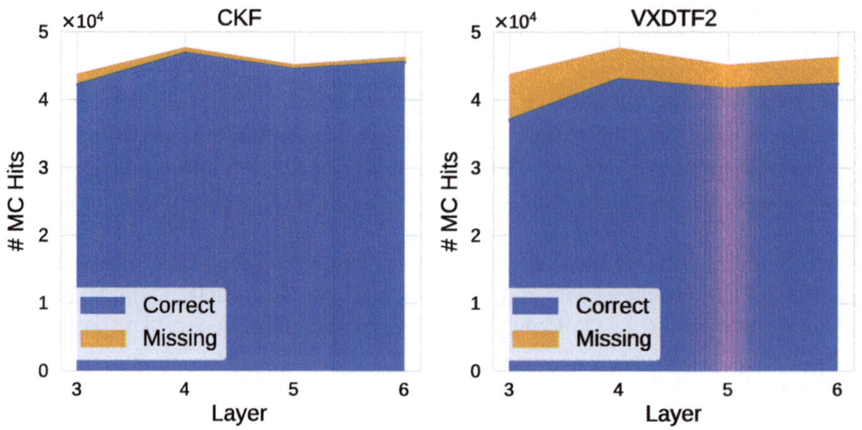

Fig. 6.21 Number of missing hits of the described CKF method in comparison with the standalone VXDTF2 with MC matching applied, only evaluated on tracks which are found by both methods. The CKF implementation is able to attach 7% more correct hits to the tracks while retaining approximately the same purity. Many VXDTF2 tracks miss a single hit, especially often in the first SVD layers, which is crucial for determining the correct track parameters

also the CKF has problems in finding the correct hit here, the total efficiency is strongly increased.

In more than 35% of the cases, the newly developed CKF method is able to attach more correct hits to the CDC tracks, in 61.4% of the cases both methods give the same result. The VXDTF2 algorithm is not tuned for secondary particles, whereas the CKF method does not include any assumption on the particles origin directly.[4] For this reason, the CKF method finds correct SVD hits in $29.6 \pm 3.2\%$ more secondary particles than the VXDTF2 method. The difference is not larger, as most of the secondary particles still originate approximately from the IP.

In conclusion, the described CKF method leads to a higher number of correctly assigned SVD hits to CDC tracks although the VXDTF2 method uses MC information for merging and has therefore even a higher performance compared to what is achievable on data.

6.6.3 Performance on ROI Finding

As discussed in Sect. 2.2, an important step of the track finder running on the HLT is to supply precise tracking information in a reasonable runtime. Equally important is

[4]To restrict the number of random combinations, broad cuts are applied to the candidates as e.g. shown in Table 6.3, which still allow for a large phase space region of secondary particles. VXDTF2 includes the assumption that particles originate from the IP when training the sectormap as well as during the Cellular Automaton traversal [15].

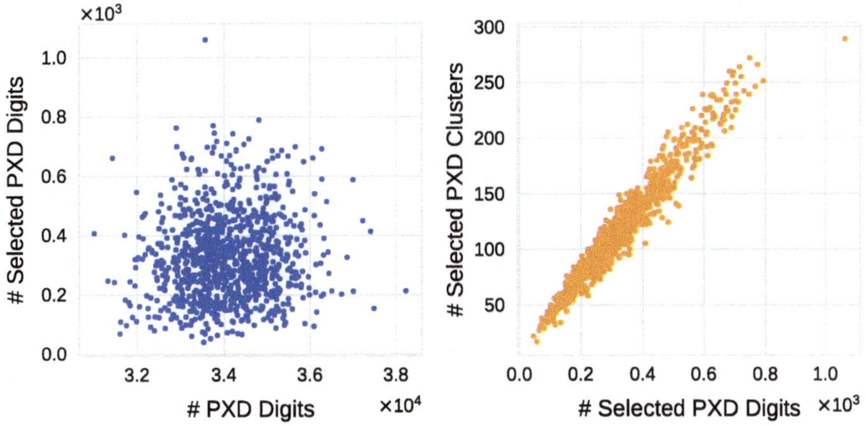

Fig. 6.22 Number of simulated and selected PXD digits including beam-induced background for typical $B\overline{B}$ events and the number of clusters formed out of those. As expected, there is no visible correlation between simulated and selected digits, as the number of digits from background is much larger as the ones from signal. In contrast, clusters are typically formed out of three PXD digits creating a strong linear dependency. As a PXD cluster already includes three-dimensional information, the usable space points for tracking are given by the number of PXD clusters

to decrease the number of PXD hits by defining regions of interest (ROIs). otherwise, the PXD cannot be read out. As this is done using tracking information from CDC and SVD, an increased precision has also a visible impact on the efficiency of the PXD cluster selection.

For the PXD selection, a list of rectangular ROIs on each PXD sensor is defined using the reconstructed SVD and CDC tracks extrapolated into the PXD volume. Only those PXD measurements that lie in one of the ROIs are transferred to the final event building and storage. As this step is performed before the reconstruction, the ROI selection uses the raw PXD digit information. A concise overview of this selection is shown in Fig. 6.22.

Compared to signal particles, background processes deposit a much larger number of digits in the PXD detector. As the number of selected digits should only depend on the correct number of hits, it is uncorrelated to the total number of digits, which is also visible in the figure. After the digit selection using the ROIs, PXD clusters are formed out of neighboring digits. Those clusters include a three-dimensional information and are directly usable as space points for the track finding discussed in the next section. The number of clusters after selection is in the order of 100–200 clusters per event, which makes it comparable to the number of space points in the SVD track finding—although there are only two PXD layers. As the input for all following algorithms are PXD clusters, only those will be counted in the following.

Figure 6.23 shows the number of selected and missing PXD clusters after the ROI selection using tracks from different algorithms. The selection prior to this thesis was using only tracks produced by the VXDTF2 without CDC track finding and merging because of technical reasons. A first small improvement was already achieved by

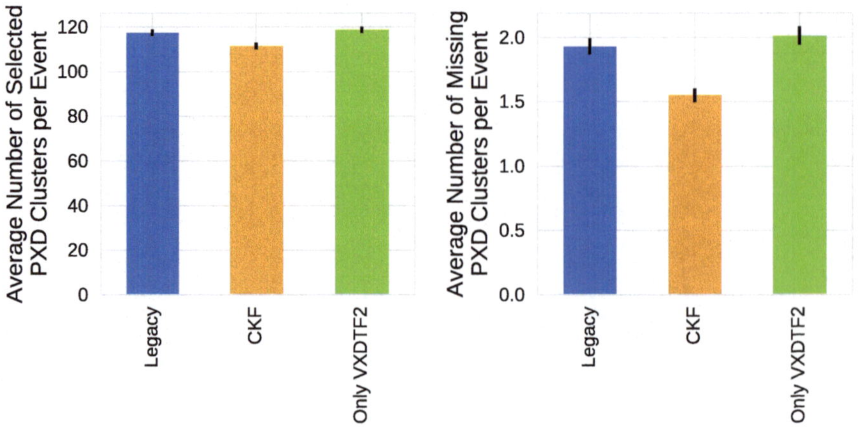

Fig. 6.23 Average number of selected and missing PXD clusters after the selection with ROIs produced with different track finder algorithms. As the new CKF implementation with additional VXDTF2 tracks gives more precise tracks, the efficiency and purity in selecting the clusters increased. The ROI selection was and is still not optimal as about 10% of the correct PXD hits are lost

using combined SVD-CDC tracks in the legacy implementation. Using the more precise and pure CKF tracks for extrapolation into the PXD decreased not only the number of missing PXD clusters per event by 0.514 ± 0.111 ($21.27 \pm 4.08\%$). It also decreased the number of total selected PXD clusters, which is very important for complying with the required bandwidth for the PXD data transport from the detector to the storage system. The efficiency of the selected PXD clusters increased from 89.4 ± 0.3 to $91.9 \pm 0.3\%$ and the purity from 16.8 ± 0.1 to $17.3 \pm 0.3\%$.

Although improved, the efficiency for selecting the correct PXD clusters is still relatively low compared to the other efficiencies seen in tracking. The ROI selection is currently optimized for the start of the Phase 3 data taking and goes beyond the scope of this thesis.

6.7 CKF for PXD Hit Attachment

The implementation of the CKF to attach PXD clusters to already found tracks follows the same principles as the SVD CKF described before. It uses the results of the new CDC and SVD track finder and the selected PXD clusters as input. As the level of background is very high in the PXD and since there are only two layers, there is no algorithm to supply additional independent tracks. This means no merging procedure is needed and the algorithm consists only of the first part including extrapolation, Kalman update and MVA filters.

Before this work, PXD hit finding was the task of the VXDTF2. As no precise fit information is used during the hit assignment (but only later in the track selection) for

Table 6.12 Concise summary of the PXD hit attachment performance using the CKF method. As expected, the small number of hits and the high amount of background compared to the SVD reduces the precision of the algorithm. This leads to a higher impurity and a lower hit efficiency. Still, the algorithm is able to find a large fraction of the PXD hits correctly which will lead to an increased resolution (see below). The numbers are normalized to all properly found and fitted input tracks

In percentage	PXD CKF (on found and fitted tracks)
Any correct PXD hits	93.17 ± 0.19
PXD hit efficiency	89.34 ± 0.20
With ≥ 1 found PXD hit	93.23 ± 0.14
Hit purity	96.65 ± 0.13

computational speed reasons, its performance in the highly dense PXD environment suffers from the high amount of background. Measures for improving the SVD hit efficiency and purity were implemented in the VXDTF2 which are at the expense of the PXD assignment, rendering this algorithm useless for PXD assignment in its current form. For this reasons, there is no other result shown for comparison in this section.

Due to the ROI finding and the small number of layers, it is sufficient to only use a rough pre-filtering based on the distance between track and hits on each layer during the CKF. The combinations are restricted to only the the nearest ten PXD clusters. Additional cuts on the distance between cluster position and helix of 0.2 cm on layer 1 or 1 cm on layer 2 increase the purity of the selected hits.

In contrast to the SVD CKF, no MVA is utilized during the extrapolation or Kalman update, as the total number of candidates is limited due to the smaller number of layers. The decision whether to attach a given PXD hit to a track or not is better taken at the end, when the full track candidate is available. This final selection is again performed using the already described BDT trained for regression using similar variables, as the summed χ^2 of the single hits or the residuals between track and clusters. A new feature only usable in the PXD is the shape likelihoods of the added clusters. The likelihood describes the probability that the measured shape was produced by a track with the given incident angles with respect to the sensor. Its calculation is calibrated on simulated events by the PXD detector experts [16]. A detailed discussion of the single building blocks as in the sections before is not repeated here, but only a concise summary of the expected performance with the nominal Belle II setup is shown in Table 6.12. The numbers in the table as well as in all shown figures in the following are normalized to the PXD clusters after the ROI selection. The PXD CKF completes the new track finding chain discussed in Fig. 6.2.

Compared to the SVD CKF, the algorithm is not able to deliver a comparable efficiency and purity. Apart from the higher amount of beam-induced background per sensor due to the smaller distance to the IP, there are also only a smaller number of PXD layers. This means the track helix parameters are not as over-determined as in the SVD case and distinguishing correct and wrong hits becomes harder.

Hit Efficiency and Second PXD CKF Iteration This can also be seen in the hit efficiency distributions for different numbers of MC PXD hits shown in Fig. 6.24. For a typical case with two PXD clusters, the hit efficiency is above 93%. However the average is $89.34 \pm 0.20\%$. Tracks with only a single cluster are found with a much higher inefficiency—also because in many cases the algorithm picked up a wrong hit (c.f. the discussion of the hit purity below). If a track deposits energy in a region of the PXD detector with overlaps, the number of hits can be larger than two. Although the algorithm does not include any special cut for this, the final multivariate selection favors tracks without overlaps. Consistently, this leads to hit efficiencies of 2/3 for tracks with a single overlap and 2/4 for tracks with two overlaps. The reason can be found in the training procedure of the MVA selection method: The final selection method is trained to select correct tracks with high hit efficiency. The track without an additional hit on the same layer resembles already well enough a correct final track—although they do not have the highest possible hit efficiency. The additional hit on the overlap layer can instead e.g. lead to a worsened average residual. Although it is possible to tune the input parameters of the training to favor longer tracks with overlaps, this leads to possibly higher impurity. As the additional hit does not influence the resolution of the tracking parameters much, finding only one hit per layer is sufficient enough.

To increase the efficiency for tracks with a single PXD hit, the CKF is applied twice: The first iteration only allows for tracks without an unoccupied layer. The second iteration operates on the remaining tracks without assigned PXD hits and tries to add additional single hits from layer 1 or 2. The beneficial influence of this second step is also shown in the Fig. 6.24.

Influence of the SVD and CDC Track Finding In Fig. 6.24 the hit efficiency for different transverse momenta is shown. Below approximately 0.2 GeV the efficiency drops rapidly from its high value of 90%. Additional to the larger influence of material effects and stochastic multiple scattering in the low-momentum regime, most of the seed tracks only have SVD hits attached, which decreases their resolution. To underline this point, Fig. 6.25 shows the hit efficiency for different categories of tracks: Tracks which have both CDC and SVD hits have a hit efficiency near 94 %, whereas tracks with only SVD or CDC hits have a much lower efficiency. The lower number of hits and the missing link between the different detectors decrease the resolution of the tracking parameters. A comparison is however difficult as e.g. CDC tracks where no SVD hit was attached have probably a worse than average resolution or are secondary tracks.

Influence of the Number of PXD Hits In Fig. 6.26, the found MC tracks are categorized by their PXD hit content. A large fraction of MC particles does not have any PXD hit assigned, caused by inefficient ROI selection or due to the fact that the particles stem from secondary decays. Compared to tracks with hits in both the first and the second layer, tracks with only hits in a single layer play a subordinate role.

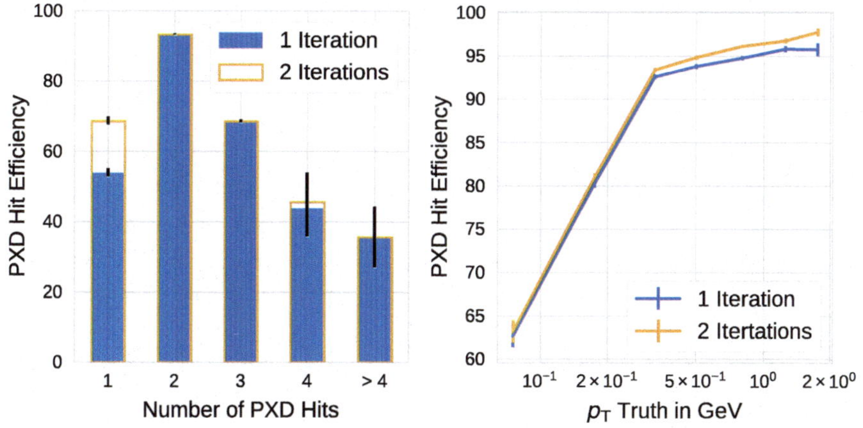

Fig. 6.24 PXD hit efficiency of all matched tracks distributed over the number of MC PXD clusters and the truth transverse momentum. The second iteration with relaxed limits on the number of missing hits increases the hit efficiency for tracks with only a single cluster—but only for higher momenta. For very low transverse momentum, multiple scattering and the worse finding efficiency of the previous track finders influence the PXD hit efficiency

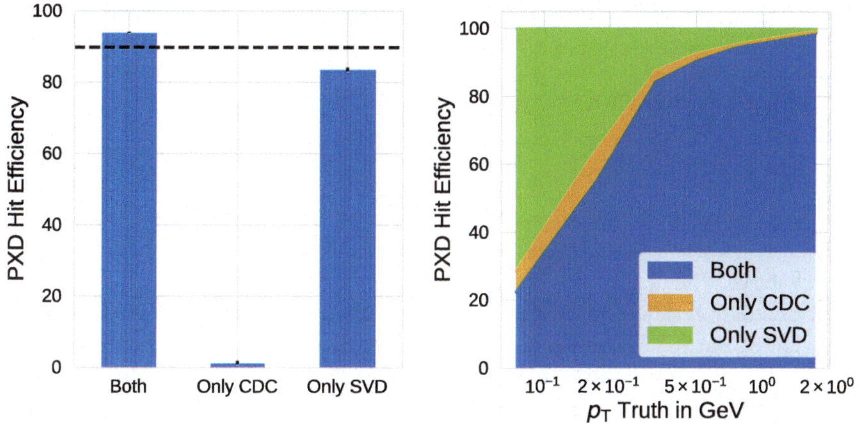

Fig. 6.25 PXD hit efficiency of all matched tracks categorized by their hit content. A lower number of hits in only a single detector is correlated to a small PXD hit efficiency, because those tracks have a worsened resolution. For lower momenta, most of the tracks only have SVD hits which leads to a smaller PXD hit efficiency in this regime. The tracks with only CDC hits in the low-momentum regime are mostly secondary tracks

As already discussed above, the algorithm fails to find all hits in cases of overlaps. In the rare case of tracks with hits in only one of the layers, the fraction of correct matches is around 60%.

Not all tracks in the very large category of particles without any true PXD clusters also have no PXD hit after the hit assignment, which decreases the hit purity.

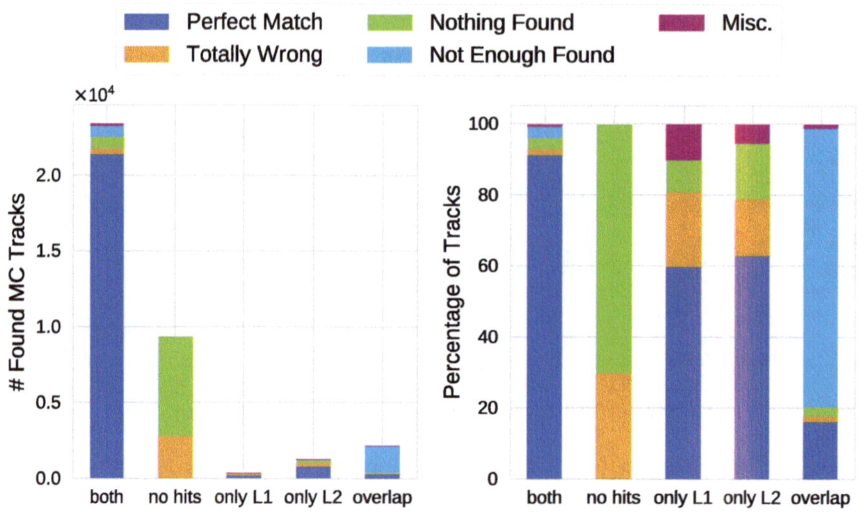

Fig. 6.26 Trackable MC tracks categorized by their PXD hit content and their status after the PXD hit assignment by the CKF. Tracks in the category "both" have a single hit on both PXD layers—with more than one hit per layer they are in the category "overlap". L1 and L2 stand for layer 1 and layer 2, respectively. The right plot is normalized to the number of tracks in each category. The uncertainty is omitted for better visibility, but especially the L1 and L2 category suffer from a low statistics. The "misc." state includes e.g. tracks where the algorithm has found all correct hits and some additional incorrect ones

Figure 6.27 shows the number of wrong assigned PXD clusters for different number of assigned clusters in total. If only a single cluster is assigned, it is wrong in about 50% of the cases, which demonstrates the difficulties of the algorithm already discussed above. The majority of tracks in this category have only a single cluster also in the simulation. The CKF finds a cluster, but it was the wrong one. As the spatial precision of the PXD is much higher than in other detector, there is no additional information to validate the assignment of this single hit using information from the other detectors and there is no other PXD hit left for further information. Although the assignment of single clusters is associated with such a large error, the additional correctly assigned clusters increase the resolution more than the wrong hits decrease it, leading to an overall benefit.

The group of tracks with two PXD clusters is very pure—apart from a small fraction where both PXD hits are incorrect. As the number of PXD clusters is very small relative to the total number of hits per track, any additional PXD hit does not influence the MC matching state of the track. If an input track from the SVD or CDC tracking algorithm is a fake, an additional added PXD hit is also classified as wrong, which is the case in more than 56% of the tracks in the green category. As the number of cases with a single wrong cluster is much smaller and the average PXD hit efficiency without taking fake input tracks into account is above 96%, it is questionable if the added hits are really wrong in all cases or are just misclassified

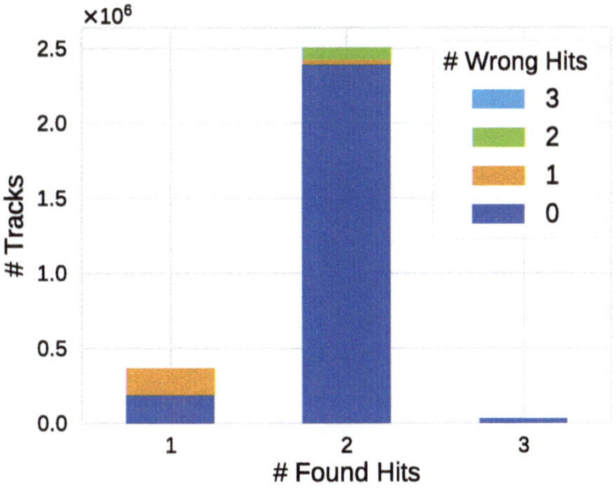

Fig. 6.27 Number of wrong hits found by the PXD CKF classified by the total number of attached PXD hits. As described, the category with only a single PXD cluster is particularly difficult. Most of the cases of two wrong attached PXD clusters come from tracks, where already the input seed was classified as fake and any additional PXD hit is therefore wrong

caused by a particularity of the MC matching procedure. As already discussed in the last sections, decays in flight play a crucial role in this case which blurs the simple decision between right and wrong MC match.

Track Parameter Resolutions In contrast to the track finding in the SVD and the CDC, which have the task to increase the finding efficiency or dismiss fakes and hit-impurities by over-determining the track in different detectors, the role of the PXD is to increase the spatial resolutions in d_0 and z_0. Figure 6.28 shows the resolutions with and without applying the PXD hit assignment with the CKF. The additional added hits have a clear positive impact on the spatial resolutions. On the momentum resolutions, no improvement was expected. As the efficiency is best for higher transverse momenta, the already improved resolution after the SVD CKF is again increased by more than 50% in these cases. In total, the new implementations of the SVD track finding as well as the implementation of the PXD hit assignment increased the resolutions of all track parameters by 60–80 % over a large momentum range.

6.7.1 Influence on $B^- \to D^0\pi^-$ Decays

As a study for a later performed physics analyses, the newly developed tracking chain was applied to study the decay of $B^- \to D^0\pi^-$, where the D^0 meson decays via $D^0 \to K^-\pi^+$. The charge conjugated decays are implicitly included. The idea of this study was not to extract physical meaningful quantities, but to demonstrate the influence of the improved efficiency and vertex resolution. The decay $B^- \to D^0(\to K^-\pi^+)\pi^-$ was chosen as it is a frequent background in many physics analyses caused by its relatively high branching fraction. Additionally, all the final decay products are charged making the reconstruction only dependent on tracking.

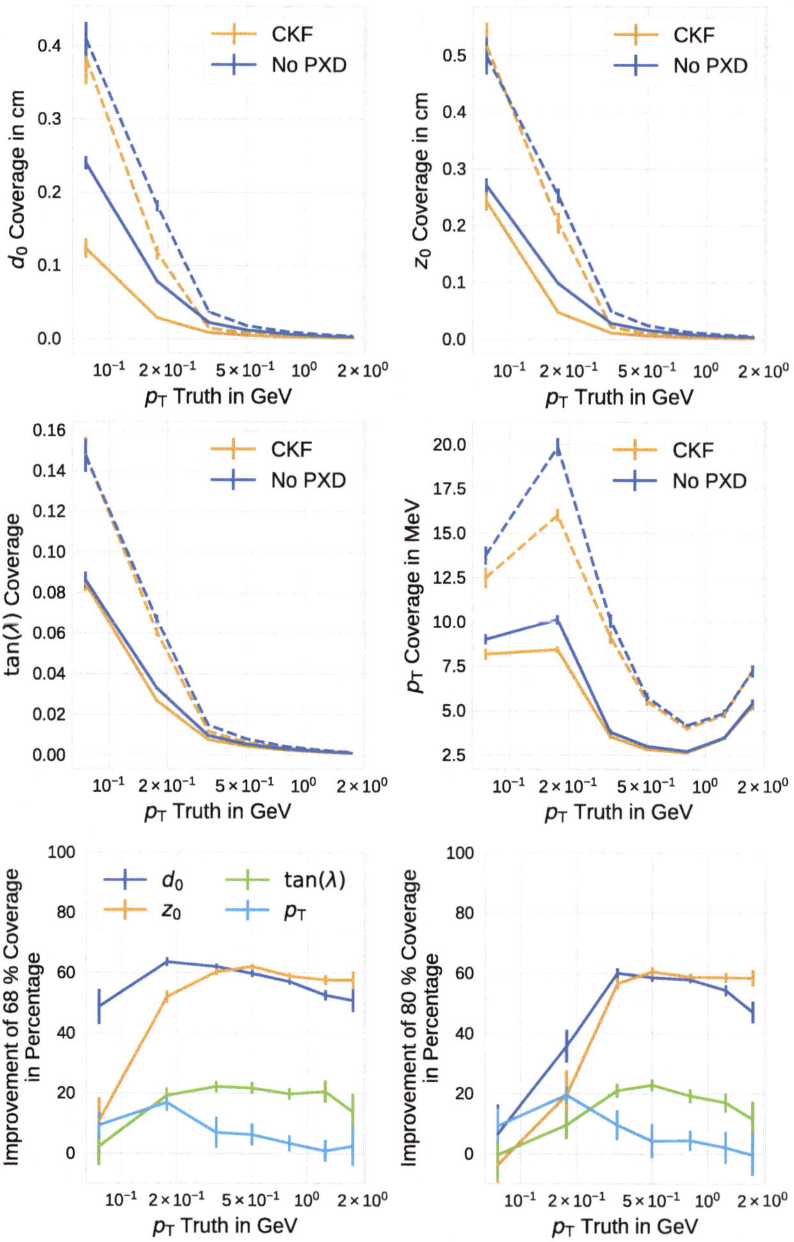

Fig. 6.28 Track parameter resolution with attached PXD hits for different transverse momenta. Shown is again the 68% and the 80% coverage of the distributions. For comparison, the resolutions after the described SVD CKF with additional VXDTF2 application are shown (c.f. Fig. 6.17) labeled as *No PXD*. The additional PXD hits improve the resolutions by more than 50% in z_0 and d_0 for higher transverse momenta. Track parameter resolution with attached PXD hits for different transverse momenta

Table 6.13 Efficiency for selecting at least a single correct B candidate in the sample of $B^- \to D^0(\to K^-\pi^+)\pi^-$ decays for different tracking algorithms. No cuts are applied at this stage. The second column shows the calculated track finding efficiency normalized to the MC track finder. It is calculated under the assumption, that the B efficiency only depends on the tracking efficiency for each particle in the decay

In percentage	B efficiency	Track finding efficiency
	Without cuts	Normalized to acceptance
MC track finding	81.83 ± 0.12	100.0 ± 0.0
SVD CKF, VXDTF2 and PXD	76.62 ± 0.14	97.83 ± 0.08
SVD CKF, VXDTF2	76.54 ± 0.14	97.79 ± 0.08
Legacy and PXD	75.41 ± 0.14	97.31 ± 0.08
Legacy	75.37 ± 0.13	97.29 ± 0.07

For the study, 10 000 $\Upsilon(4S)$ events were simulated with beam-induced background and reconstructed with the described tracking algorithms extended by the remaining post-tracking reconstruction, e.g. ECL cluster matching and PID extraction. In the sample, one of the B mesons out of the $\Upsilon(4S)$ decay always decays into the above mentioned signal channel, the other with a $b \to c$ quark decay. As the legacy implementation does not include any PXD track finding, the new CKF implementation is used on top of it to have a fair comparison.

At first, each fitted track was used as both a possible K or π candidate and combined to a D^0 candidate, which then was combined with another π to a B candidate. Using the MC matching procedure described in [17] for the combined particles, each B candidate is classified as either correct or wrong. The first column in Table 6.13 shows the efficiency to find at least one correct B candidate in an event for different tracking algorithms.

To interpret these numbers correctly, they are normalized to the expectation. Although the detector is built to cover as much phase space of the decaying products as possible, the acceptance is not 100% and therefore also the expected efficiency is below 100%. As three tracks need to be reconstructed for the decay, the overall efficiency using the MC track finder of $81.83 \pm 0.12\%$ means the acceptance is 93.54 $\pm$ 0.05% for each track, assuming[5] an equally distributed efficiency. As the particles all originate from near the interaction point and have relatively high momenta, the anticipated track finding efficiency is high (c.f. Table 6.11). The efficiencies normalized to a single track and the acceptance of the detector by dividing with the efficiency for the MC track finding algorithm are also shown in the second column of the table. As seen before, the newly developed algorithm has a slightly higher track finding efficiency, as the VXDTF2 is executed on a cleaned hit sample. The small improvement leads to an improved efficiency of $1.87 \pm 0.82\%$ for the B signal

[5]This assumption is clearly wrong, as the three decay products have different kinematic properties. The shown averaged efficiency is therefore only a rough estimate.

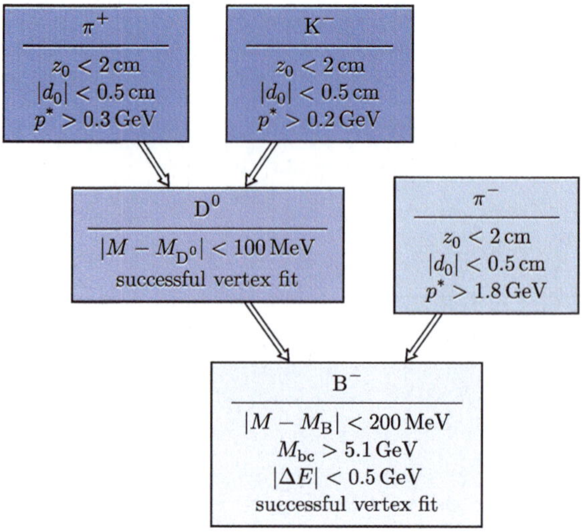

Fig. 6.29 Simple cut-based selection criteria on the kinematic properties of the reconstructed particles. The selection criteria are loosely motivated by the physical properties of the decay and are not optimized for best performance but should only simulate a typical physics analysis. The variables follow the notation introduced in [18]

candidate before cuts. As no cuts are applied, there is no difference between using the PXD hits or not seen.

In a typical analyses, a signal selection is performed for improving the purity and reducing the number of wrong combinations. As this is not the topic of this thesis, only a rough cut-based selection without optimization of the efficiencies is applied. The selection criteria are shown in Fig. 6.29 and are loosely motivated by the kinematic distributions of the particles. The variables follow the notation introduced in [18]. A later physics analysis would need to optimize the selection, e.g. by utilizing a multivariate method.

The efficiency to find at least one correct B candidate with all its decay products in the event after the selection is given in Table 6.14 and is reduced by approximately 6–10% compared to Table 6.13. The cuts are chosen in such a way to increase the purity of candidates to above 80%. As the generated event sample includes only the decay under study, the remaining wrong candidates are solely created by combining the decay products of the signal B in a wrong way or by combining particles from both B mesons in the event.

The difference between the legacy implementation and the described newly implemented CKF algorithm with VXDTF2 is increased with the selection cuts. The selection with the new tracking is $3.63 \pm 0.22\%$ points better than the legacy reconstruction. In decay channels with a more complicated event topology and particles with more challenging kinematic properties, the difference can be even higher. Even the relatively small improvement of $3.63 \pm 0.22\%$ points can already have a very large impact on the final selection.

The increased resolution of track parameters has a direct influence on the properties of the particles and more correct candidates pass the same cuts. This analysis was performed without taking other B decay into account that could mimic the signal

Table 6.14 Efficiency for finding at least a single correct candidate in the simulated signal sample per event for different track algorithms. The cuts discussed in Fig. 6.29 are applied as signal selection. The second and third column show the fraction of events, where more than one correct candidate is found before and after the selection

In percentage	B efficiency	Double rate	
	After selection	Without cuts	After selection
MC track finding	74.39 ± 0.13	–	–
SVD CKF, VXDTF2 and PXD	68.69 ± 0.15	1.19 ± 0.04	0.033 ± 0.006
SVD CKF, VXDTF2	67.34 ± 0.14	1.23 ± 0.03	0.030 ± 0.005
Legacy and PXD	66.19 ± 0.15	4.38 ± 0.07	0.105 ± 0.011
Legacy	65.05 ± 0.16	4.49 ± 0.07	0.148 ± 0.011

decay. It is however immediately clear that a higher resolution on the properties of the particles also leads to a better suppression of the background. Although especially studies of CP violations benefit from a good spatial resolution of PXD hits, also in this toy analysis the efficiency increased by another $1.35 \pm 0.21\%$ when including PXD hits.

The remaining columns in the table above show the fraction of events, where more than a single correct candidate is found. This can only happen if one of the particles is reconstructed twice by the tracking algorithms. As expected from the decreased clone rate of the newly implemented SVD algorithm, the fraction of those events drops by a large number. The event selection removes the majority of the cases, but some clones remain, possibly deteriorating the measurement. Compared to the legacy implementation, those cases are crucially reduced by the new implementation.

To show the increased resolution, Fig. 6.30 shows the standard deviation of the reconstructed z and r vertex position of the correct B mesons compared to their MC values. The increased resolution on the tracking parameters is passed along to the vertex resolutions, although an additional vertex fit is performed afterwards. The new implementation with the SVD CKF and the VXDTF2 has again a twice as good resolution in both important quantities compared to the legacy reconstruction. When comparing to the legacy implementation without PXD, which was the default before this thesis, the resolutions were enhanced by 45% and 57%, respectively.

6.7.2 Reduced PXD Setup

The final focus for the described tracking methods is to increase the track reconstruction performance for the final data taking period, as most of the integrated luminosity will be collected with this nominal detector setup. During the construction of the PXD sensors, unexpected delays prevent the timely installation of the final PXD setup before the start of Phase 3 in 2019. As a countermeasure, Phase 3

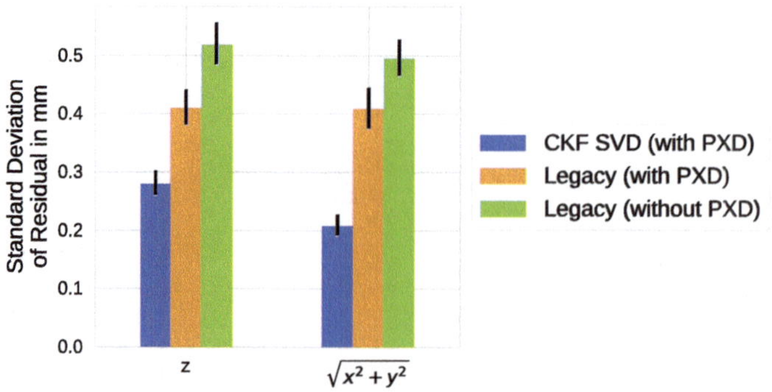

Fig. 6.30 Standard deviation of the $r = \sqrt{x^2 + y^2}$ and z parameter of the reconstructed vertex position of the signal B using different track finders. As expected from the resolution studies before, the properties of the particles are known with much more precision using the new implementation. As especially the z component is crucial for many time-dependent CP violation measurements, the improvement by roughly a factor of two is very beneficial

will start with a reduced PXD setup. Only the innermost layer 1 will be installed and a small number of distinct sensors in layer 2 [19]. Still, it is planned to include the full PXD in the nominal detector setup. Although the fraction of data, which will be recorded with this reduced PXD setup, will be comparable small to the full data set, it is nevertheless crucial to study the performance of the hit assignment in this environment. The final instantaneous luminosity which will be delivered by the accelerator during this period is not known yet. The shown studies will therefore be performed using the background anticipated for the final instantaneous luminosity, although this highly overestimates the background level. Only the first layer of the PXD will be used in the studies as the acceptance region of the small number of sensors in layer 2 is small.

Having a single PXD layer poses major problems to the track reconstruction algorithms, as no additional precise hit measurements can be used to increase the assignment purity. As seen above, the purity and efficiency of the algorithm was the worst in cases with one single PXD cluster. Additionally, the distance between the innermost SVD layer and the outermost PXD layer is now increased making the extrapolated distance und the uncertainty after the extrapolation larger. The multivariate methods used in the PXD hit assignment were retrained to cope with these facts. The remaining SVD and CDC tracking are left unchanged. The impact of the missing layer 2 on the performance of the algorithm is shown in Table 6.15.

The efficiencies and purities drop to the same level, which the two-layer PXD CKF has for single-clustered tracks. The reasons for this have already been discussed in the previous sections.

Figure 6.31 shows the change in resolution of the tracking parameters between the nominal Phase 3 setup and the reduced PXD detector. Also shown are the parameters before the application of the PXD hit assignment. As the tracking algorithm for the CDC and SVD detectors and the detector setup for those outer detectors are

Table 6.15 Comparison of the PXD hit assignment performance in the CKF algorithm with and without reduced PXD setup. As expected, the missing layer 2 of the detector decreases the efficiency as well as the purity by a large fraction. A reasonable assignment is possible in more than 84% of the cases making the PXD adding valuable information to the tracks

In percentage	PXD CKF (L1 + L2)	PXD CKF (L1 only)
	(on Found and Fitted Tracks)	
Any correct PXD hit	93.17 ± 0.19	84.41 ± 0.28
PXD hit efficiency	89.34 ± 0.20	82.61 ± 0.30
With $\geq$1 found PXD hit	93.23 ± 0.14	88.39 ± 0.24
Hit purity	96.65 ± 0.13	90.32 ± 0.24

unchanged, the helix parameter resolutions are unchanged without any PXD hit assignment.

When PXD hits are added, the picture is somewhat different. Both quantities d_0 and $\tan(\lambda)$ deteriorate by approximately 20% and z_0 by 50% up to 100% for transverse momenta around 200 MeV. At those momenta, many tracks are only found in the SVD decreasing the hit assignment performance as seen in the subsections before.

With the problems of only a single PXD cluster per track and the larger extrapolation distance, the chance of increasing the performance of the PXD assignment in the reduced PXD setup are small. Although the resolutions are deteriorating without the second PXD layer, they are still better with using the PXD and can be used for physics analyses with the collected data. It is planned, that the period with the reduced setup spans only a small fraction of the full data taking period and that much of the data will be recorded with the full precise PXD setup, where the presented algorithm performs best.

6.8 Performance in Phase 2

The studies presented so far in this chapter were performed using the simulation of the final Belle II detector setup as planned for Phase 3 data taking—with or without the second PXD layer. In summer 2018, the full detector except most of the VXD sensors was already assembled and tested with collisions at low instantaneous luminosity. This first runs allowed to test the hardware and the full readout chain and for first physics analyses using recorded data. More than 200 pb^{-1} were recorded [20] and are currently analyzed. Although the main focus was to test the hard- and software, the recorded data can also be used for interesting physics analyses, e.g. searches for dark matter candidates.

Although the VXD was replaced by a single slice of lower quality VXD sensors and measuring devices for the background rates, it is still possible to perform track extrapolation into the VXD volume and test the performance on the first recorded data.

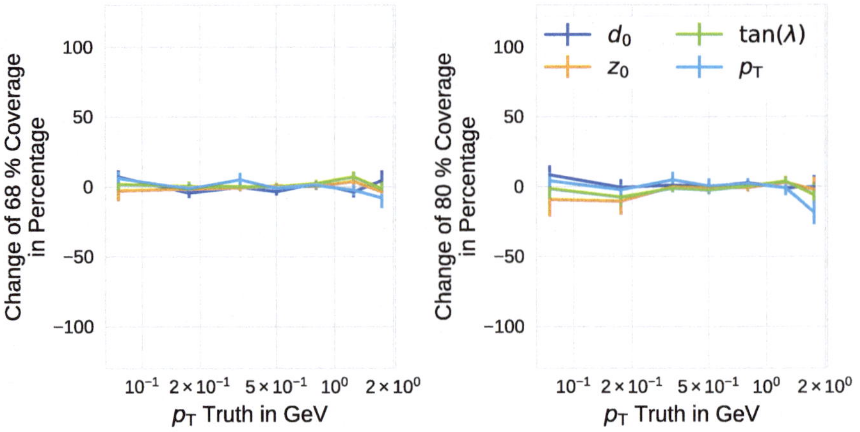

(a) Without attached PXD hits

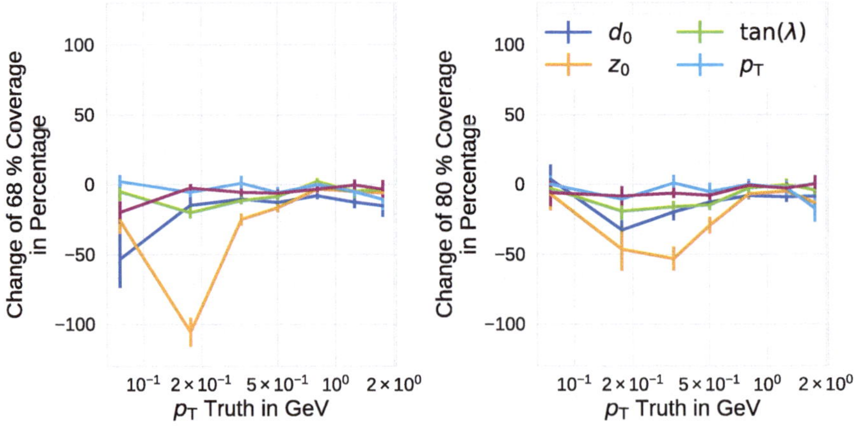

(b) With attached PXD hits

Fig. 6.31 Comparison of the resolution distributions with and without the full PXD setup. The resolutions in both detector setups are subtracted and normalized to the nominal PXD setup. A negative results means, the reduced setup gives worse resolutions. When no PXD hits are added, no change in the resolution is seen, as expected. The final resolution with attached PXD hits however deteriorates especially for z_0, because of less correctly added pixel hits

6.8.1 Characteristics of the Phase 2 Setup

For this work, the only important difference between the described final Phase 3 setup and the Phase 2 setup used for recording are the number and arrangement of the VXD sensors. Figure 6.32 shows the setup used during the data taking period in 2018.

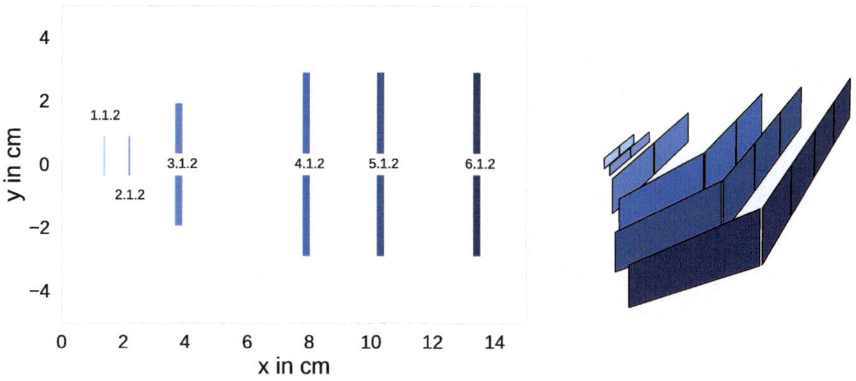

Fig. 6.32 VXD Setup used during data taking in Phase 2 in x-y and three-dimensional view. Only a single ladder in the r-ϕ plane of each PXD and SVD layer is used. Also the arrangement of the sensors is different from the Phase 3 geometry (cf. Fig. 2.3). In z direction (not shown here), the sensor setup is the same as in Phase 3

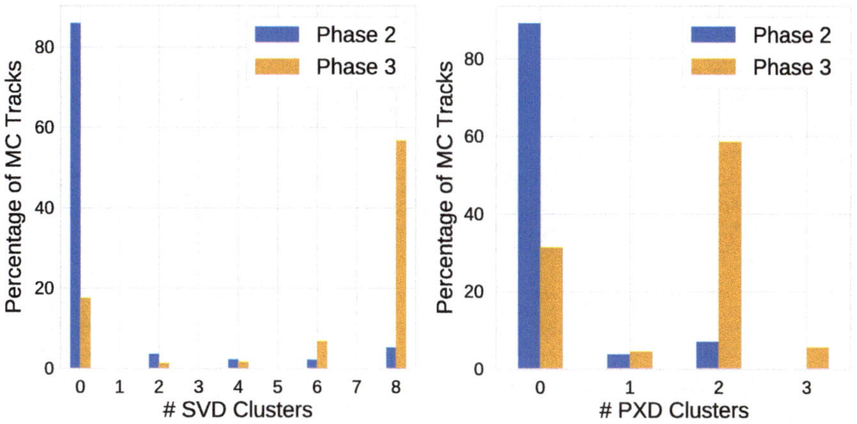

Fig. 6.33 Expected number of clusters in the vertex detectors per track in comparison between Phase 2 and Phase 3 geometry. As the acceptance is much smaller in Phase 2, the relative fraction of tracks without or with only a small number of VXD clusters is much higher

Figure 6.33 shows the number of SVD and PXD hits per track expected from MC simulations in comparison between Phase 2 and Phase 3. The acceptance of the vertex detectors is smaller and the limited number of VXD sensors decreases the number of tracks which have PXD or SVD hits. Especially the number of tracks with hits from more than one vertex detector is smaller. As seen in the previous sections, tracks with a small number of hits pose major difficulties to the tracking algorithms, as the track is not over-determined anymore by additional hits.

On the other hand, the instantaneous luminosity of the recorded data is orders of magnitude lower than the peak luminosity of Phase 3, which was used for the studies

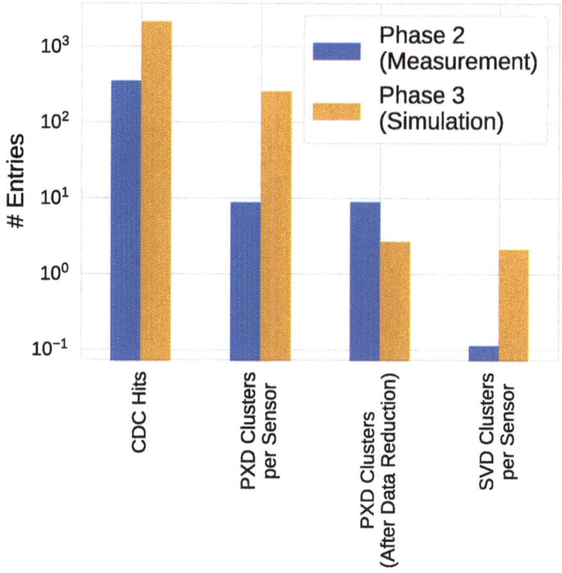

Fig. 6.34 Simulated and measured background distributions for Phase 2 and Phase 3 respectively. The background for Phase 2 was recorded in runs 4650, 4651 and 4666 with an instantaneous luminosity between 1.3 and $1.7 \times 10^{33}\,\mathrm{cm}^{-2}\,\mathrm{s}^{-1}$ using random triggers. The Phase 3 simulation assumes the estimated full instantaneous luminosity of $8 \times 10^{35}\,\mathrm{cm}^{-2}\,\mathrm{s}^{-1}$. As expected, this has a large impact on the number of expected background hits in each detector. For Phase 2, all PXD clusters without data reduction are used. For PXD and SVD, the number of background clusters per sensor is shown to make Phase 2 and Phase 3 comparable

presented to far. This has also a strong impact on the beam-induced background expected in the detector. Figure 6.34 shows a comparison between measured and simulated beam-induced background for Phase 2 and Phase 3. The lower number of expected hits per PXD sensors and the lower number of sensors itself make it possible, to not use any data reduction during the Phase 2 recording. The following Phase 2 results will therefore use all PXD digits without ROI selection.

In Phase 2, the number of hits per track is smaller making the distinction between wrong and correct candidates harder and the extrapolation distances are larger increasing the uncertainty. Still, as shown in the following, the CKF is able to reconstruct SVD and PXD hits in the Phase 2 geometry with reasonable efficiency. Although useful for studying the tracking efficiency, the detector hardware performance and alignment, the VXD information plays only a subordinate role during physics analyses of the recorded events, as the number of tracks which include VXD information is limited.

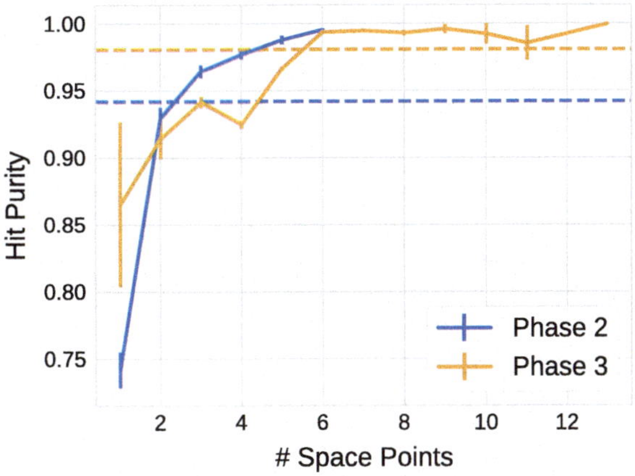

Fig. 6.35 VXD hit purity over the number of attached space points for Phase 2 and Phase 3. The beam-induced background is smaller leading to a higher purity in Phase 2 for the same number of space points (except for a single space point, which is somewhat special). But as the fraction of tracks with a lower number of hits is smaller in Phase 3, the average purity is increased (shown as dashed line)

6.8.2 Performance on Simulated Events with the Phase 2 Geometry

In Table 6.16, important figures of merit of the tracking algorithms in Phase 2 with beam-induced background taken from random triggered events are compared to their corresponding values in Phase 3 with simulated full luminosity background.

The overall finding efficiency is lower in Phase 2 compared to Phase 3. Tracks with a small number of hits in the CDC cannot be found by the CDC algorithm or are not fittable without proper z information. The VXDTF2 is able to recover these tracks in Phase 3 but not in Phase 2 because most of those tracks do not travel through the small SVD sensor volume. The hit purity and efficiency of the CDC algorithm benefit from the lower number of background hits in the CDC. This also improves the fake and clone rate. As expected, the performance of the VXD track finding is not as good as with Phase 3 geometry due to multiple reasons. There are more tracks with only one or two SVD space point or PXD clusters, which leads to a worsened hit purity and in turn also to a decreased efficiency as was already pointed out in the last sections. Especially, there are tracks which do not transverse the higher layers of the SVD but only one or two of the inner layers. The extrapolated distance between the CDC seed and these layers is larger which increases the uncertainty. Additionally, the precise SVD measurements in between are missing which help to estimate the track parameters correctly. This is also visible in the hit purity over the number of added space points shown in Fig. 6.35 for Phase 2 and Phase 3. Because of the higher amount of background, the hit purity for lower numbers of space points is worse in Phase 3. The total amount of tracks with such a low number of hits is however strongly reduced with the full VXD setup, so those impure tracks do not influence the overall hit purity much (shown as dashed line).

Table 6.16 Basic figures of merit for the described tracking algorithms consisting of CDC track finding, CKF for SVD and PXD and additional VXDTF2 with CKF merging for Phase 2 and Phase 3. Due to the lower number of background hits, the CDC track finding is improved in Phase 2 with higher purity and hit efficiency. In the VXD, the algorithm is still able to find tracks with reasonable efficiency and purity despite the geometry is more difficult. The overall finding efficiency is decreased, as tracks with only a small number of hits in the CDC cannot be recovered with the VXDTF2 as in Phase 3

In percent	Phase 2	Phase 3
Finding efficiency	79.39 ± 0.20	82.80 ± 0.25
Primaries	89.12 ± 0.17	96.03 ± 0.12
Fake rate	2.26 ± 0.08	6.49 ± 0.15
Clone rate	3.96 ± 0.11	6.26 ± 0.16
PXD hit efficiency	84.4 ± 0.6	89.57 ± 0.19
SVD hit efficiency	88.92 ± 0.35	91.33 ± 0.15
CDC hit efficiency	87.24 ± 0.14	69.96 ± 0.24
PXD hit purity	90.5 ± 0.5	92.55 ± 0.18
SVD hit purity	94.99 ± 0.24	98.99 ± 0.05
CDC hit purity	94.52 ± 0.03	89.98 ± 0.05

6.8.3 *Performance on Phase 2 Data*

The data collected in runs 4957, 4958, 4963, 4964, 4675, 4678, 4679 and 4680 are compared to simulated $B\overline{B}$ events using measured beam-induced background. There was no HLT cut applied during the data taking. As the implemented L1 trigger during this period did not select the events with a high purity, an offline event selection is performed. Only tracks with reconstructed parameters of $|d_0| < 0.5\,\text{cm}$ and $|z_0| < 0.5\,\text{cm}$ are taken into account to dismiss tracks from beam-induced background. Additionally, only events with at least 5 of these selected tracks are used in the study. Still, the event topologies are different between simulated and recorded events. Preliminary calibration and alignment constants are applied in the reconstruction of the data. The final integrated luminosity is not known on a per run level yet, so the absolute number of tracks or hits in the data samples is not comparable.

The simulated events have been generated with a shift of the interaction point to resemble the data. Figure 6.36 shows the reconstructed d_0 over ϕ_0 of the selected tracks. Using the helix parametrization of the trajectory, the interaction vertex position (x_y, v_y) is given by

$$\begin{aligned} v_x &= d_0 \sin\,\phi_0 + d_0 \cos\,\phi_0 \\ v_y &= -d_0 \cos\,\phi_0 + d_0 \sin\,\phi_0 \end{aligned} \Rightarrow d_0 = v_x \sin\,\phi_0 - v_y \cos\,\phi_0\,,$$

so the reconstructed d_0 changes with respect to ϕ_0 for vertex position other than (0, 0). Using the recorded data, the vertex position for this run is extracted to be

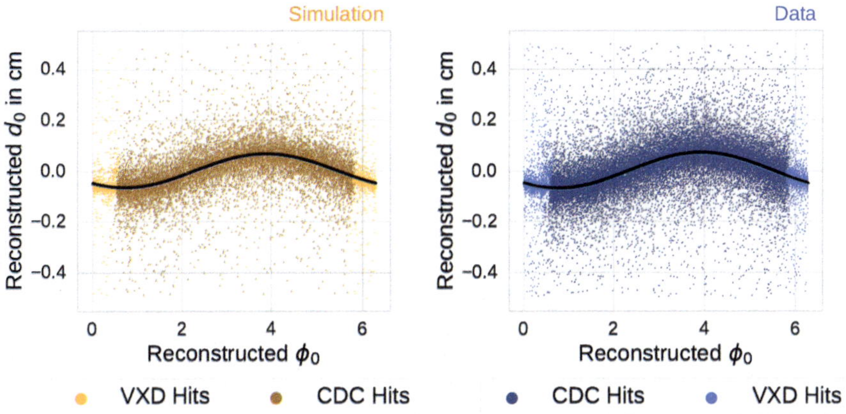

Fig. 6.36 Reconstructed d_0 and ϕ_0 of simulated and recorded data. Due to the relationship between d_0 and ϕ_0, the vertex position can be extracted. The simulated data is also produced with this offset. Shown are all selected tracks separated if they have only CDC hits or also VXD hits. As the VXD only covers a small area in ϕ, the tracks can only have VXD hits in this region

$$(v_x, v_y) = (-0.048, 0.049)\,\text{cm}\ ,$$

which is used for simulation (also shown in the figure).

As the particles are produced homogeneously in ϕ, each track has the same chance in hitting the VXD detectors. This allows to compare simulation with recorded data although the track multiplicity and the luminosity is different. Figure 6.37 shows the number of recorded and found hits per selected track for each VXD sensor. A hit is hereby given by a PXD or SVD cluster. As the level of background is small and the acceptance of the VXD limited, the number of expected hits from simulation is already small per track. On data however, only a fraction of these hits are recorded. There are multiple reasons for this, e.g. inefficiencies of the sensors, problems with the readout or wrong calibration and alignment constants. A discussion of these effects goes beyond the scope of this thesis.

As there are typically two SVD clusters per layer (u- and v- strip), the ratio of found hits in the SVD is higher than in the PXD. As the ϕ coverage drops with higher radii, the number of hits also decreases. Additionally, sensors in the central region of the detector (e.g. sensor 2 on layer 4) are hit more often by a track and also the reconstruction efficiency is higher in this region. Both simulation and data follow these overall expectations.

As a summary, Fig. 6.38a and b show the fraction of tracks with CDC, SVD or PXD hits and the number of attached VXD space points. The algorithm is able to attach PXD or SVD hits to approximately the same percentage of tracks in data and simulation. However, it has a bias towards smaller number of attached space points on data. This is in line with the hit inefficiencies before the track finding described above. The VXDTF2 is the only track finding that is able to produce tracks without

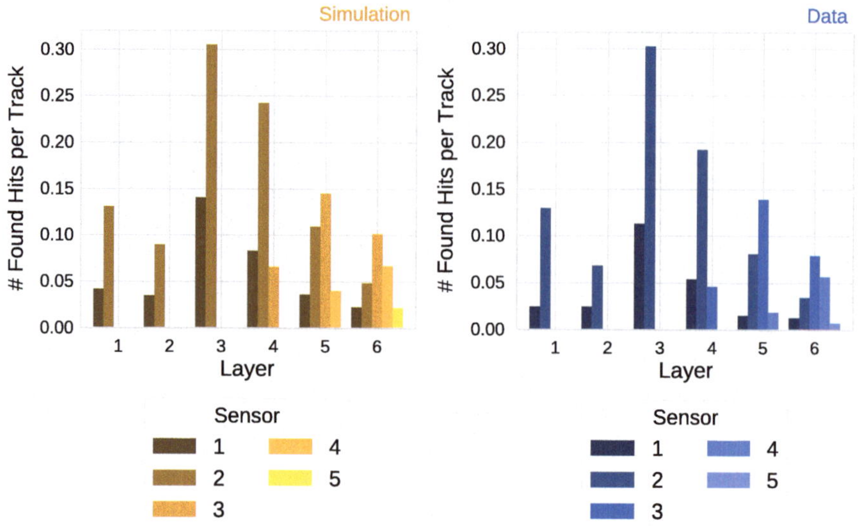

Fig. 6.37 Used hits (SVD or PXD cluster) per VXD sensor for simulation and recorded data. The numbers are normalized to the number of found tracks to compare the different luminosity and multiplicity in the samples. The visible differences in the recorded data are caused by misalignment, inefficiencies in the detector or improper calibrations

CDC hits. As discussed before, it requires a minimum number of three hits in the SVD to find a track. The hit inefficiencies seen already before track finding effect this track finder the most, leading to a large deviation between data and simulation in the category of tracks without CDC hits. Additionally, it may be effected by the displaced vertex position.

Apart from the number of hits, also the track parameters calculated on data are of interest. Figure 6.39 shows the reconstructed z_0 and d_0 separately for tracks with only CDC hits or with at least a single VXD hit. As expected from the sections before, the resolution on these crucial spatial tracking parameters increases significantly with attached vertex detector information both on simulation and data.

The figure show two interesting properties. First, the d_0 peak of the tracks with VXD hits is shifted towards negative values both in simulation and data. This is due to the displaced vertex in the x-y plane and the limited acceptance of the VXD detectors in the ϕ angle. As seen in Fig. 6.36, the average of the reconstructed d_0 of al tracks is close to zero, but the small area of the VXD leads to a finite bias due to the dependency between d_0 and ϕ_0. This is also visible in the simulation as the same displaced vertex was chosen during generation. Second, the z_0 resolution seems to be worse on data compared to simulation. The truth value of the z vertex position is however not known and as seen in Fig. 6.40 also not determinable reliably from data. The z vertex was tuned for testing by the accelerator experts during Phase 2 and the final setup with high precision in the IP position was not reached. It is therefore unclear, if the track parameter resolution is worse or the interaction point was shifted.

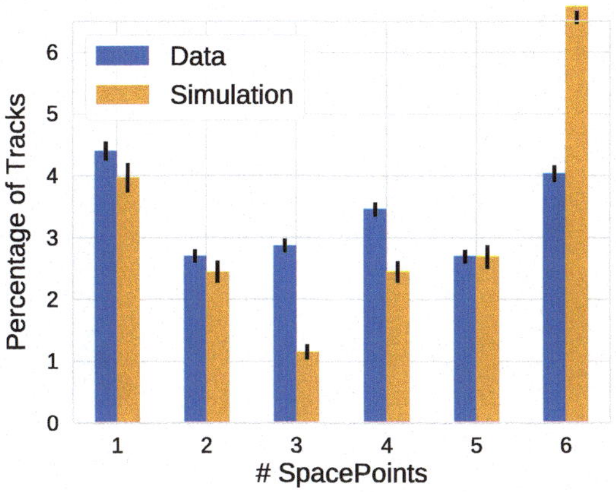

(a) Space points in total.

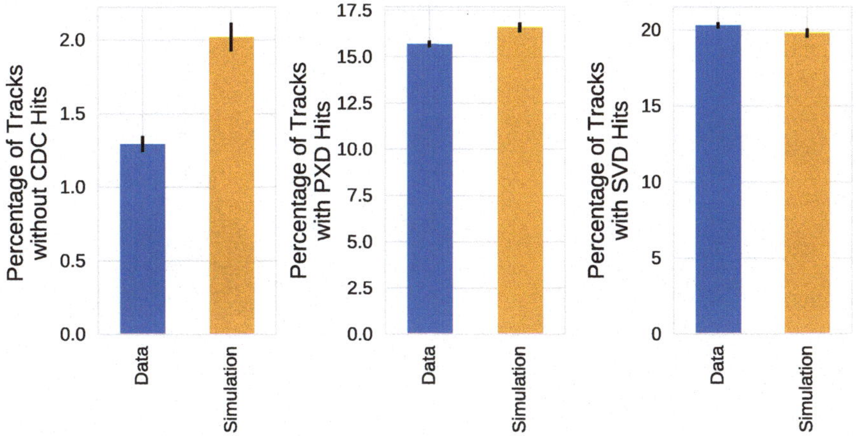

(b) Fraction of tracks with hits in each subdetector.

Fig. 6.38 Number of attached space points to tracks by any of the VXD track finders in comparison between Phase 2 simulations and recorded data. Although there are visible differences between simulation and data especially for a high number of space points, it is also clear that the algorithms are able to attach hits using recorded events

Figure 6.41 compares the median unbiased residual between reconstructed track and attached hit for different VXD layers. The unbiased residual is determined by performing the track fit without the hit and comparing the track position with hit measurement in the same plane. The figures show a very good agreement between simulation and recorded data in the sub-µm range. The remaining deviations especially in the SVD could be caused by misalignment.

In conclusion, the implemented track finding algorithms developed in this thesis behave similarly well on simulation as well as on recorded data during Phase 2. The

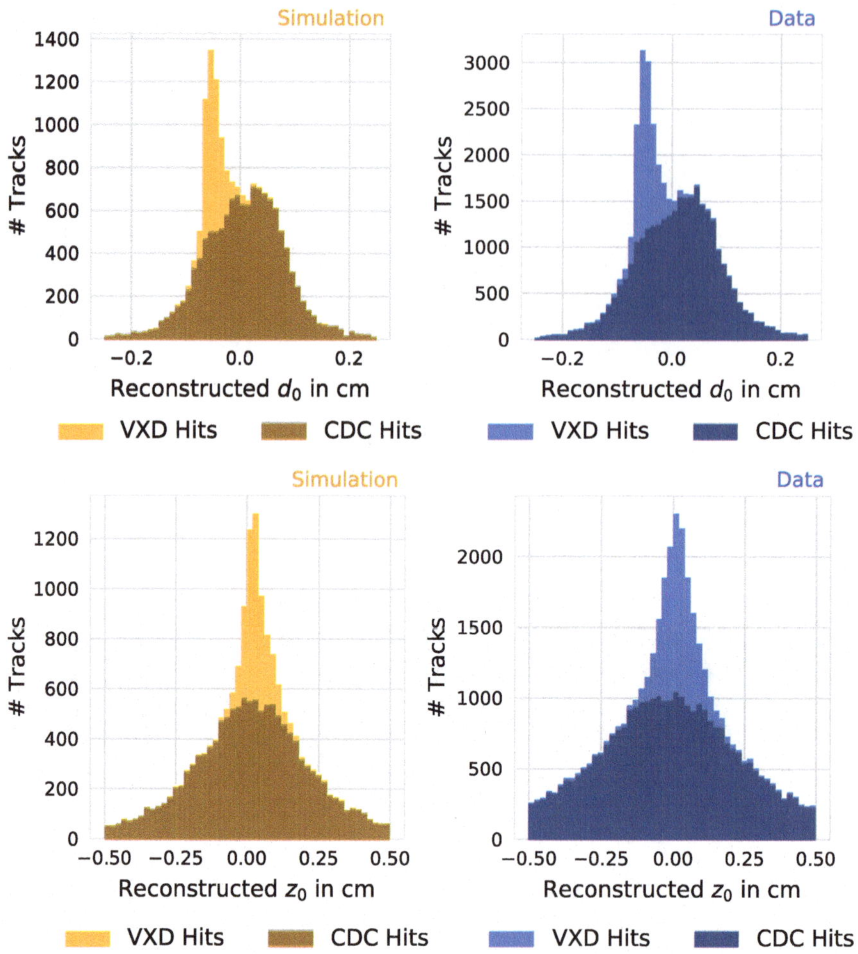

Fig. 6.39 Reconstructed d_0 and z_0 for simulation and data separated between tracks, which have only CDC hits (in dark color) or with at least a single VXD hit (in light color). As expected, the tracking parameter resolution is increased by a large amount with the additional VXD hits. The shifted peak in the d_0 distribution originates from the displaced vertex and the limited ϕ acceptance of the VXD detectors. The z_0 resolution in data is worse probably due to an displaced interaction point in the same direction

robustness against missing hits or a small number of attached VXD hits make the CKF a crucial tool especially when large hit inefficiencies happen. A large fraction of SVD hits as well as all PXD hits in the Phase 2 recording can only be correctly attached with the newly developed algorithm.

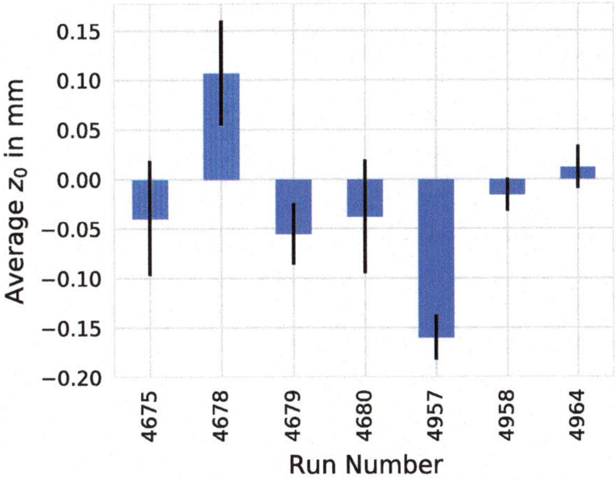

Fig. 6.40 Average reconstructed z_0 for different runs in the used data sample. The quantity is comparable to the z position of the IP and is not very constant between different run periods. It is however unclear, if the reconstruction has a bad resolution or of the accelerator was not tuned correctly

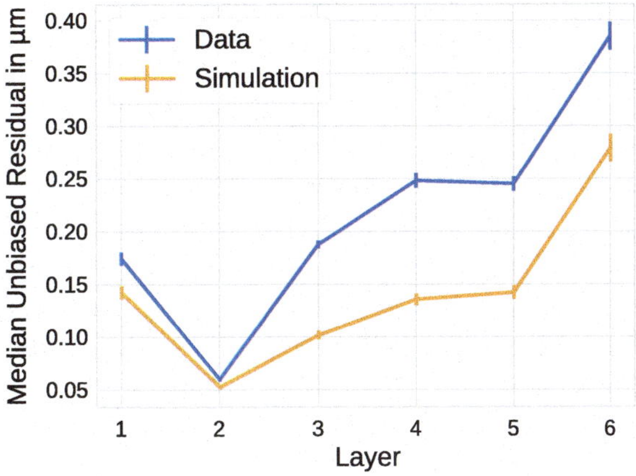

Fig. 6.41 Median unbiased residual between attached hits and reconstructed tracks for simulation and recorded data on different VXD layers. The differences between simulation and data is below 1 µm and could be caused by a misalignment of the sensors

6.9 Summary

This chapter presented the first implementation of a combinatorial Kalman filter (CKF) for the Belle II experiment. The algorithm is used for SVD hit finding, SVD and CDC track merging as well as for PXD hit assignment in the standard reconstruction used for simulation and during data taking. The usage of the new algorithms increased the finding efficiency and decreased the number of clones significantly. A positive impact was seen in the hit efficiencies and especially in the resolutions of the tracking parameters, which will be directly visible in any analysis performed at Belle II. Additionally, it is currently the only algorithm capable of using the precise spatial measurements of the PXD despite high beam-induced background. The implementation was validated with first recorded data from Phase 2.

The framework which was used for the implementation is generic. The basic parts—the extrapolation and Kalman update, the algorithmic representations of the detector measurements and the track state as well as the filters and the tree search—are independent of the detector setup or their task. This turned out to be already valuable for implementing the SVD, PXD as well as the merger CKF. However, there are additional possibilities how the CKF can be also used beneficially at Belle II. Each of the possibilities need to be studied thoroughly and are only discussed here for reference for future work.

CDC with SVD seeds The most evident application for an additional CKF in the default reconstruction is to add CDC hits to already found SVD tracks. The VXDTF2 has a high finding efficiency and—although not as precise as the SVD CKF—produced track candidates with good parameter resolution. Compared to the vertex detectors, the CDC has a smaller level of background and a much larger number of layers. This makes it easier to decide between wrongly and correctly reconstructed candidates during the tree search. Although the application is very near to the ones presented in this thesis, completely different problems need to be faced in the CDC. If restricting to a maximum of two possible hit candidates per layer, a typical high-p_T seed can easily create $2^{56} \approx 10 \times 10^{16}$ candidates to test, which is obviously way out of reach for any meaningful computation. This is why a step-wise approach, finding tracks in one superlayer at a time and advancing from superlayer to superlayer afterwards, is used. The implementation is currently developed by a group at DESY and shows already auspicious results [5].

Outer Detectors The tracking volume of the Belle II detector is surrounded by the ECL and the KLM in the barrel and endcap region (BKLM and EKLM). To combine the measurements in these detectors, the reconstructed tracks need to be extrapolated into these detector volumes and the hits or clusters need to be attached to the track. The task is similar to the PXD hit finding and is currently implemented using a similar technique based on the same extrapolation algorithm of the tracks. Using the well tested common framework of the CKF for this task is beneficial. This would make it also possible to track particles, which leave the CDC into the calorimeter, deposit energy there and reenter the CDC later, possibly with deteriorated track parameters due to energy loss.

Monopole Tracking Due to the clean environment at Belle II, it is the perfect experiment to search for specific classes of monopoles [21–23]. These hypothetical particles with magnetic charge would have a distinct signature in the detector due to their special behavior under an applied magnetic field. Especially, they cannot be found easily using standard track finding procedures. Dedicated track finding based on the Hough algorithm exists for this task [24]. The CKF does not assume a specific track model except in the extrapolation of the track state from one detector plane to the other, which could easily be adapted to the use case of monopoles.

The generic CKF implementation turns out to be useful at various places in the Belle II reconstruction. The already implemented algorithms improved the track reconstruction crucially for all following reconstruction steps as well as the physics analyses. Possible future applications can either refine the results of tracking or other reconstruction products even more, or open new possibilities for searches of interesting new physics.

References

1. Mankel R, Spiridonov A (1999) The Concurrent Track Evolution algorithm: extension for track finding in the inhomogeneous magnetic field of the HERA-B spectrometer. Nucl Instrum Methods Phys Res Sect A Accel Spectrom Detect Assoc Equip 426(2–3):268–282. http://linkinghub.elsevier.com/retrieve/pii/S0168900299000133, https://doi.org/10.1016/S0168-9002(99)00013-3
2. Mankel R (2004) Pattern recognition and event reconstruction in particle physics experiments. Rep Prog Phys 67(4):553–622. arXiv:0402039 [physics], https://doi.org/10.1088/0034-4885/67/4/R03
3. CMS Collaboration (2014) Description and performance of track and primary-vertex reconstruction with the CMS tracker. J Instrum 9(10):P10009–P10009. arXiv:1405.6569, https://doi.org/10.1088/1748-0221/9/10/P10009
4. ATLAS Collaboration (2017) Performance of the ATLAS track reconstruction algorithms in dense environments in LHC Run 2. Eur Phys J C 77(10):673. arXiv:1704.07983, https://doi.org/10.1140/epjc/s10052-017-5225-7
5. Künzel M, Glazov A, Guo A (2017) The VXD to CDC combinatorial Kalman filter. In: F2F tracking meeting. Pisa. https://indico.desy.de/indico/event/19161/session/6/contribution/9/material/slides/0.pdf
6. Braun N (2015) Momentum estimation of slow pions and improvements on the track finding in the central drift chamber for the Belle II experiment. Master's thesis, KIT. https://ekp-invenio.physik.uni-karlsruhe.de/record/48740/files/EKP-2015-00052.pdf
7. Wagner J (2017) Track finding with the silicon strip detector of the Belle II experiment. Master's thesis, KIT. https://ekp-invenio.physik.uni-karlsruhe.de/record/48930/files/EKP-2017-00057.pdf
8. Weiler MH (2016) Optimization of the VXD-CDC matching algorithm for the Belle II detector. Bachelor's thesis, KIT. https://ekp-invenio.physik.uni-karlsruhe.de/record/48873/files/EKP-2016-00044.pdf
9. Billoir P (1989) Progressive track recognition with a Kalman-like fitting procedure. Comput Phys Commun 57(1–3):390–394. http://linkinghub.elsevier.com/retrieve/pii/001046558990249X, https://doi.org/10.1016/0010-4655(89)90249-X
10. Frühwirth R (1987) Application of Kalman filtering to track and vertex fitting. Nucl Instrum Methods Phys Res Sect A Accel Spectrom Detect Assoc Equip 262(2–3):444–450. http://linkinghub.elsevier.com/retrieve/pii/0168900287908874, https://doi.org/10.1016/0168-9002(87)90887-4
11. Gamma E (1995) Design patterns: elements of reusable object-oriented software. Addison-Wesley
12. Rauch J, Schlüter T (2015) GENFIT—a generic track-fitting toolkit. J Phys Conf Ser 608:012042. arXiv:1410.3698, https://doi.org/10.1088/1742-6596/608/1/012042
13. Keck T (2017) FastBDT: a speed-optimized multivariate classification algorithm for the Belle II experiment. Comput Softw Big Sci 1(1):2. arXiv:1609.06119, https://doi.org/10.1007/s41781-017-0002-8
14. Hoecker A, Speckmayer P, Stelzer J, Therhaag J et al (2007) TMVA-Toolkit for multivariate data analysis. http://arxiv.org/abs/physics/0703039
15. Lettenbichler J (2016) Real-time pattern recognition in the central tracking detector of the Belle II experiment. PhD thesis, Technische Universität Wien. http://www.ub.tuwien.ac.at

16. Schwenker B (2017) New ideas on PXD hit reconstruction and calibration from beam data. In F2F tracking meeting. Mainz. https://indico.desy.de/indico/event/18594/session/1/contribution/2/material/slides/0.pdf
17. Pulvermacher C (2015) Analysis software and full event interpretation for the Belle II experiment. PhD thesis, Karlsruher Instituts für Technologie. https://ekp-invenio.physik.uni-karlsruhe.de/record/48741/files/EKP-2015-00053.pdf
18. Bevan AJ, Golob B, Mannel T, Prell S et al (2014) The physics of the B factories. Eur Phys J C 74(11):3026. arXiv:1406.6311, http://link.springer.com/10.1140/epjc/s10052-014-3026-9, https://doi.org/10.1140/epjc/s10052-014-3026-9
19. Niebuhr C (2018) PXD for phase 3-BPAC review october 2018. In: Belle II general meeting. Tsukuba. https://desycloud.desy.de/index.php/s/ZcYtqeL5zAt6cmo#pdfviewer
20. Shebalin V (2018) ECL luminosity monitoring. In: Belle II general meeting. Tsukuba. https://kds.kek.jp/indico/event/27570/session/50/contribution/15/material/slides/0.pdf
21. Fryberger D (2018) On magneticons and some related matters. arXiv:1412.8677, http://arxiv.org/abs/1807.09606
22. Pinfold JL (2014) The search for magnetic monopoles. Phys Today 69(10):40–46. arXiv:1412.8677, https://doi.org/10.1063/PT.3.3328
23. Sullivan MK, Fryberger D (2017) Magnetic charge search for the Belle II detector. arXiv:1707.05295, http://arxiv.org/abs/1707.05295
24. Neverov D (2018) The search for magnetic monopoles. In: AEPSHEP 2018

Chapter 7
Conclusion

With the start of Phase 3 in the beginning of 2019, Belle II will deliver its crucial part for the study of multiple interesting physics models and theories. Its planned instantaneous luminosity and its clean e^+e^- collisions may help in solving many mysteries and open questions, and will refine the parameters of the Standard Model of particle physics.

Very precise measurements with high statistical precision cannot be performed without a well-working reconstruction. Only the extraction of physically meaningful parameters from the recorded energy depositions and electrical charges in dedicated detectors makes them usable for physics analyses. Any error or inefficiency in this step has an influence on every search, every study and every measurement on the recorded data. Without high efficiency and purity in the particle reconstruction, a large fraction of physics analyses cannot be perform without extending the cost-intensive data acquisition period and the significance of all measurements is reduced.

The online reconstruction running for the high level trigger (HLT) decision has the largest influence on the success of Belle II. The HLT is the software trigger stage after the hardware-based Level 1 (L1) trigger. Any collision event dismissed at this stage is irretrievably lost for physics analyses, as it will never be stored to disk as raw data. Due to the online reconstruction happening on the HLT live with the data taking, strong timing requirements apply to the software. In this thesis, these requirements were studied and the feasibility to also run the full offline reconstruction with only small changes online was confirmed for the first time (see Fig. 7.1). For this, the full software framework for the HLT trigger decision has been written from scratch. Detailed processing time studies including possible influences of deteriorations on the expected scaling behavior for the multi-core calculation were performed. To reach the strong timing requirements, a continuous HLT trigger decision with an exemplary early Bhabha veto was introduced in this thesis. In the end, it could be confirmed that the reconstruction with the changes implemented in this thesis is able to process the incoming data stream from the L1 trigger in time. Without

N. Braun, *Combinatorial Kalman Filter and High Level Trigger Reconstruction for the Belle II Experiment*, Springer Theses,
https://doi.org/10.1007/978-3-030-24997-7_7

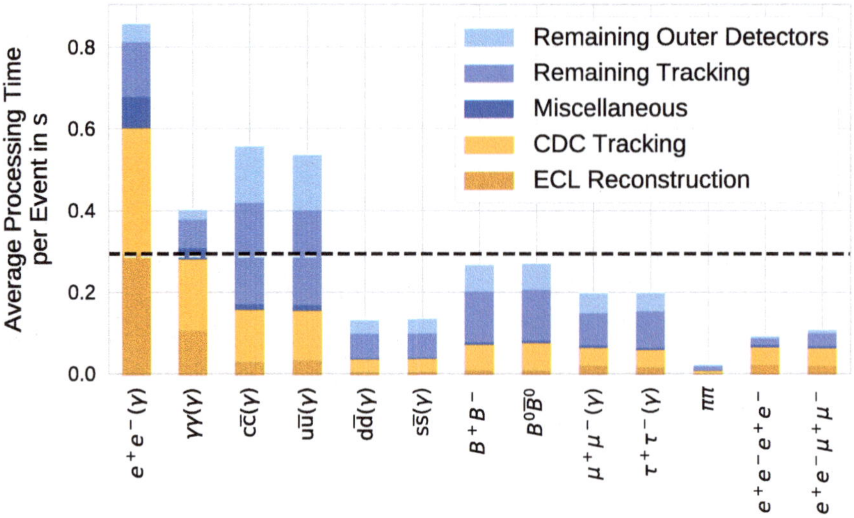

Fig. 7.1 Measured processing time including an exemplary early Bhabha veto, weighted according to the cross section of the different channels. The average reconstruction time of 0.3 s shown as a dashed line is below the required 0.32 s

the early trigger stage implemented in this thesis or the lessons learned from the processing time studies, it would not be clear if the HLT needs to dismiss collision events due to insufficiently fast processing time. A new framework developed in this thesis based on the open-source ØMQ library solves multiple issues of the legacy implementation for multiprocessing calculations. It has been shown to have no drawbacks in the processing time while including beneficial features such as an event backup or an automatic worker healing. It is prepared for future developments like network streaming or high-throughput calculations and will be used in the upcoming data acquisition of Phase 3.

To perform a reconstruction on the recorded detector data, many subdetectors need precise information on the event time T_0. Notably, the drift time calculation in the central drift chamber (CDC) or the background subtraction in the electromagnetic calorimeter (ECL) cannot be performed without knowing when exactly the collision happened. Although the acceleration revolution frequency is known, the bunch crossing leading to the recorded event still needs to be determined. In this thesis, a first implementation of the full event time extraction using the CDC information was introduced. It works better than initially expected and leads to a precision on the extracted time below the required 2 ns (see Fig. 7.2a). The implementation was successfully tested on the first recorded data of Phase 2. The CDC method implemented in this thesis is the first fully-working example of a time extraction method in the reconstruction software and is now used as a benchmark for the other algorithms. Without the time information extracted with this CDC algorithm, it would not be possible to use the data that will be recorded during Phase 3 in 2019.

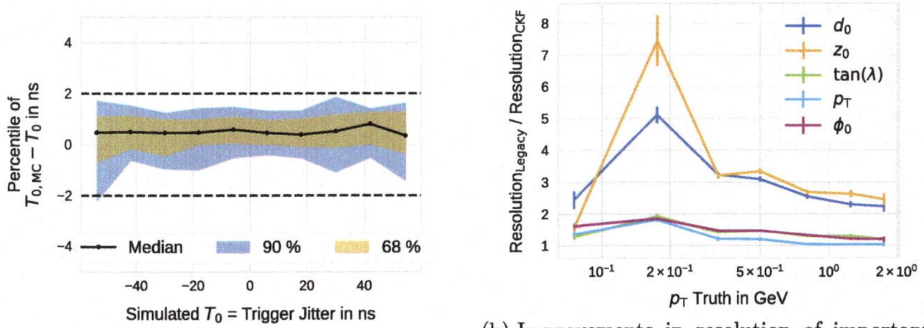

(a) Results of the combined time extraction on simulated $B\overline{B}$ events with beam-induced background. For a large fraction of cases with high trigger jitters, the resolution is better than 2 ns, which is precise enough to determine the bunch crossing reliably.

(b) Improvements in resolution of important track parameters due to the CKF implemented in this thesis. The legacy implementation without the possibility to use the precise PXD hits is compared to the new implementation with refined SVD tracking and PXD hit attachment.

Fig. 7.2 Results of the event time estimation and the CKF developed in this thesis

The largest and most influential part of the particle reconstruction is the track reconstruction. Many other detectors, e.g. the time-of-propagation counter (TOP), need precise track information to extract meaningful measurements. Additionally, tracking plays a crucial role in the majority of physics analyses, as it is the only part of the reconstruction that is able to measure all charged decay products with high precision. This work introduced a new tracking concept in Belle II: the combinatorial Kalman filter (CKF) which is now an important part of the default tracking reconstruction. With this additional tracking algorithm, the resolution in many track parameters is improved by more than a factor of two up to a factor of seven (see Fig. 7.2b). Additionally, the rate of wrongly found tracks due to missing links between the strip-vertex detector (SVD) and the CDC was nearly halved. Due to its high precision, the CKF is the only algorithm able to use the important pixel detector (PXD) as the high rate of beam-induced background processes makes it hard to find hits reliably in the subdetector. However, only the PXD enables later physics analyses to reach the vertex position precision needed for many CP-violation studies or other measurements. The newly implemented algorithm was successfully tested on simulated MC events and data recorded during early Phase 2 data acquisition. During Phase 2, its results were a significant input to the detector and alignment experts. The software framework developed for this work is general enough to allow for multiple additional use cases of the algorithm, some of which are currently under investigation and will increase the efficiency and resolution even further.

Three crucial ingredients for the success of the Belle II data taking were developed in this thesis and enable a large field of interesting physics analyses at Belle II. Without a reliable and fast HLT decision, a correct event time, and an efficient and precise track reconstruction, no accurate measurement can be performed on the recorded data.

Printed by Printforce, the Netherlands